THE DOVETAIL BOOK

Tools & Techniques for Mastering a Classic Woodworking Joint

From the Editors of *Popular Woodworking*

CEDAR LANE PRESS

CONTENTS

Page 20

1. HAND-CUT DOVETAILS

Tools for Dovetailing	6
Your First Hand-Cut Dovetails	10
Frank Klausz on Learning to Cut Dovetails	20
Dovetail Ruler Trick	28
Dovetail Dilemma	31
4 Tips for Dovetailing by Hand	34
Precise Hand-Cut Dovetails	36
Tips to Avoid Dovetail Gaps	48
Houndstooth Dovetails	50
The Art of Making Dovetailed Drawers	60
Tapered Sliding Dovetails	67
Compound-Angle Dovetails	71
Impossible Dovetails	75
Mitered Dovetail Box	82
Dovetail Station	90
Shaker Tray with a Little Embellishment	98

2. POWER-CUT DOVETAILS

Bandsawn Dovetails	108
Power-Assisted Half-Blind Dovetails	113
Router Table Dovetails	120
Sliding Dovetails with a Router	127
Make a Sliding Dovetail at the Table	131
Table Saw Dovetails	136
How to Make Condor Tails	141
Make a Drawer with a Half-Blind Jig	149
10 Tips for Using a Dovetail Jig	157
Half-Blind Dovetails by Jig	161

CONTRIBUTORS 166
MANUFACTURERS 166
INDEX 167

Page 108

1

HAND-CUT DOVETAILS

If you want to know how to cut dovetails by hand, this is the place to start. Join dovetail masters like Frank Klausz, Tom Caspar, and Christopher Schwarz as they examine the ins and outs of cutting this legendary joint by hand. If there is a method, tip, or piece of advice that works for hand-cutting dovetails, you'll likely find it here. In addition to half-blind and traditional dovetails, you'll also venture into variations like the tapered sliding dovetail, houndstooth, and more. Discover the proper tools to use, jigs you can make to simplify the process, and then try out what you've learned on a Shaker tray or mitered box.

TOOLS FOR DOVETAILING

Here are a few pointers on choosing, modifying, and using my favorite set of tools.

BY TOM CASPAR

Cutting dovetails by hand is one of the most satisfying jobs a woodworker can do. Sure, it can be frustrating, because mastering the art requires patience and experience—plus a nice set of tools.

Here's mine. I wouldn't claim that they're the best set for everyone, but they've served me well. I keep them in a box that also functions as a stand (see "Dovetail Station," p. 90).

LAYOUT TOOLS

■ *Square.* A 4" to 6" square works best because it's easy to balance. A plain engineer's square can work fine, but when you must mark the same measurement many times over, a square with a sliding blade and rule is handier.

■ *Sliding T-bevel.* This tool is for laying out dovetail angles, of course. Some people prefer small T-bevels, but I prefer a large one. A large bevel allows me to draw long layout lines, which are easier to sight along than short ones.

■ *Saddle square.* This tool is a nice luxury—it allows you to draw a line across the end of a board and a mating line down its face at the same time. The two lines will always meet.

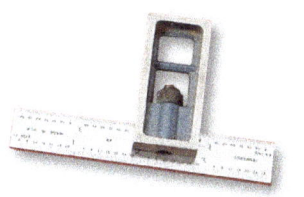
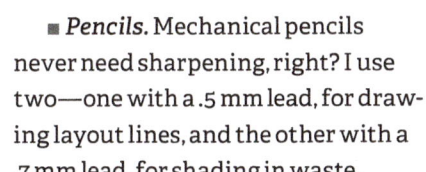

Of course, you can draw these two lines in separate steps using a regular square, but sometimes you'll make a mistake and they won't line up. A saddle square takes the risk out of the operation.

■ *Cutting gauge.* This tool is used for scoring the baselines of a dovetail joint across the grain. When sharp, it should make a line as fine as one made by a surgeon's knife. Most cutting gauges come with knives that are sharpened to a point. I modified mine to have a round profile by rotating it along its axis on a grinder. A round profile will stay sharp much longer because it has more points of contact with the wood. A point only has one.

■ *Striking knife.* This tool is used for laying out very fine lines. I prefer a knife with a flat side and a beveled side, rather than two beveled sides (like a utility knife). Bearing the flat side against a square or a dovetail makes it easier to draw an accurate, unwavering line.

■ *Pencils.* Mechanical pencils never need sharpening, right? I use two—one with a .5 mm lead, for drawing layout lines, and the other with a .7 mm lead, for shading in waste.

CUTTING AND CHOPPING TOOLS

■ *Dozuki saw.* A Japanese-style pull saw takes a little getting used to, but it's an awesome tool. It requires very little effort to cut, so you can concentrate on following a line rather than fighting the saw. I've got nothing against a good Western-style saw, which cuts on the push stroke, but it will require occasional sharpening and setting. Both operations can be quite difficult. Most Japanese saws don't need to be sharpened—ever. When a blade gets dull, you replace it with a new one. Japanese saws come with short or long blades; I prefer a long blade because it requires fewer strokes to cut to the same depth. Each stroke introduces the possibility of error, so the fewer strokes you make, the straighter your cut will be.

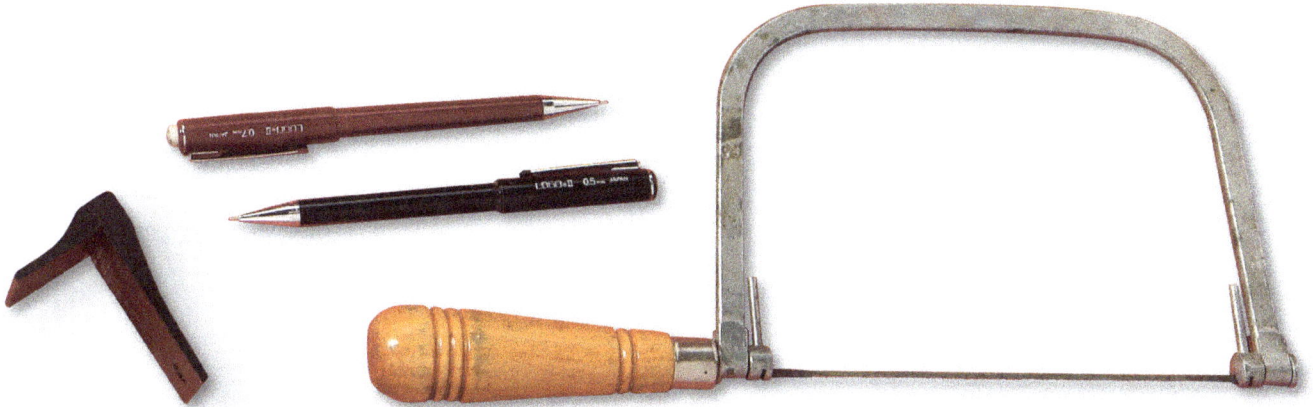

HAND-CUT DOVETAILS | Tools for Dovetailing

- *Coping saw.* This saw is used to remove the bulk of the waste from a joint before the remainder is chopped away. A fancy model certainly isn't necessary, but you should use a high-quality blade with the appropriate number of teeth for the work at hand. I prefer a blade with 15 teeth per inch—not too fine, but not too aggressive, either.

- *Chisels.* I use two sets of chisels. One set has square sides, while the other has sides that taper to a sharp point. I use the first set for paring and hone them at 25°. (A low angle makes a chisel easier to push.) I use the second set for chopping and hone them at 30°. (The steeper the angle, the longer an edge will last.) The tapered edges of the second set allow me to get into angled corners, so I rarely have to use a skew chisel to clean out a dovetail. The tapers are angled at 12° and run back about ¾". I created the tapers by using a grinder.

- *Mallet.* I prefer a round mallet to one with a square head. Both will work fine, of course, but you have to pay more attention to how you hold a square mallet to avoid a glancing blow. A round mallet is more forgiving. I like a mallet with some heft—about 16 to 20 oz. The extra weight means you don't have to strike a chisel with so much force. Just dropping a heavier mallet on a chisel often does the job.

- *Strop.* Stropping a chisel renews its edge in just a few seconds. Stropping is easiest when a chisel's edge is hollow-ground—you just balance the edge on heel and toe and go for it. I hone my dovetail chisels the same way, without a jig, to make them easier to strop.

- *Thin blades.* I use these blades for paring the sides of skinny sockets. One of the blades is just a block-plane iron; I made the other from a broken-off power hacksaw blade, wrapping tape around one end to make it more comfortable to hold.

- *Shims.* A set of playing cards is good for more than just a game of poker. Years ago, while in business, I made dovetails by the artisan's quick method of sawing and chopping to a line. These days, I slow down. I saw and chop away from a line, then pare to a line using a guide block and shims. The block is clamped right on the line. I place a few shims against the block and remove one after each shaving. My rule of thumb: The thinner the shaving, the more accurate the paring. ■

Tools for Dovetailing | **HAND-CUT DOVETAILS**

JOINT SURVIVORS

BY ADAM GODET

This anecdote will give you the gumption you need to choose and learn how to use the correct tools for dovetailing—you never know what test your joints will have to survive.

On a cold, rainy day in December, I was returning home from running errands in my Washington, D.C., neighborhood. I was listening to NPR, and thinking about the leftover pizza I was going to eat for lunch before heading into the shop.

Then, at an intersection two blocks from my house, I found myself stopped, pointed 90° in the wrong direction, static coming from the radio, the hat I'd been wearing on the floor of the car, and airbags deployed.

I'd been in an accident. One of us—we both thought the other guy—had run a red light. It was never determined who made the error; both cars were totaled but all humans were, more or less, OK.

The night before the wreck, my wife, Jen, and I had been at a holiday craft show. It was a fun night that ended late. When we got home, rather than unpack the car, I left a large pine box (not that kind) filled with cutting boards in the trunk, along with various other sundries and detritus. While I stood beside my wrinkled Honda Civic waiting for the tow trucks and police to arrive, the rain and temperature both falling, I regretted this moment of laziness. Fortunately, a friend down the street was able to come to the scene with Jen to empty the car before it was towed.

Hours later, after the adrenaline had passed and the headache and fogginess of a mild concussion had settled in, Jen mentioned that the box holding the cutting boards had broken.

Thus chastened, I began thinking about my new country-western song: "My car is gone, my holiday plans are hosed, my head hurts, I can't see straight, and my dovetails broke." Then I paused and thought, "Where did it break?

What broke? The wood or the joint?" I got dressed and went back out through the rain to the shop where the box was sitting.

The joy I felt when seeing that the wood had broken and not the joint was the consolation I needed.

Both cars were going about 30 to 35 miles per hour; the box was about 36" long x 12" wide, and was filled with wooden cutting boards. There was a lot of force on that box, and I would expect something to give—especially considering that the box was made from ¾" pine boards. But my joints didn't break. The wood near the joints did, yes—but the dovetails held.

In a moment when not a lot of things felt good or certain—my health, my driving capabilities, my holiday vacation, my car, etc.—I had craftsmanship.

I knew I could still cut tight dovetails that could stand up to more reasonable force than any bookshelf should ever face. And while I was told woodworking with the fogginess of a concussion was a bad idea—advice I heeded—I found myself standing in my shop with my tools, feeling a gentle calmness and clarity in the wake of calamity.

YOUR FIRST HAND-CUT DOVETAILS

The right techniques and tools will give you a good start on mastering this fine traditional joint.

BY LONNIE BIRD

Dovetails have long been recognized as the premier joint for casework and drawers—and for good reason. They're the strongest way to join the corners of a box, and they look great.

However, dovetails also have a reputation as a difficult joint to master. But cutting dovetails by hand only looks difficult. It's actually just a process of sawing and chiseling to a line. It's that easy. (And with a bit of practice, everyone can saw and chisel to a line.) In fact, when I teach dovetailing, I start people out not by cutting dovetails, but just sawing to a line. Once you've mastered sawing to a line, you're on your way to creating this time-honored joint.

No doubt you've seen the multitude of jigs available for routing dovetails. But there are several good reasons for skipping the jigs and learning to cut dovetails with hand tools. Undoubtedly the main reason is the pleasure that comes when crafting the joint with a saw, chisel, and mallet. Cutting dovetails is fun. Another reason is the personal satisfaction of meeting the challenge head-on. And once you develop the skills, you'll find that you can cut a variety of dovetail joints that can't be produced with a jig. Keep reading, and I'll show you step by step how to lay out and cut woodworking's most beautiful joint.

A FEW TERMS

Before diving in, it's helpful to understand some of the terms associated with dovetails. All dovetails have two mating parts: tails and pins. Tails are usually wider than pins and are tapered on the face. Pins are narrow and tapered on the ends. It's the tapered, mechanical interlock, combined with the long-grain gluing surfaces, that give dovetail joints their tremendous strength.

Through-dovetails are the most common type; the joint is aptly named because each member of the joint goes "through" the adjacent member. Consequently, through-dovetails can be viewed from either face.

Half-blind dovetails can only be viewed from one face; on the adjacent face the joint is hidden. On a typical drawer, through-dovetails are used to join the side pieces to the back and half-blind dovetails join the sides to the drawer front.

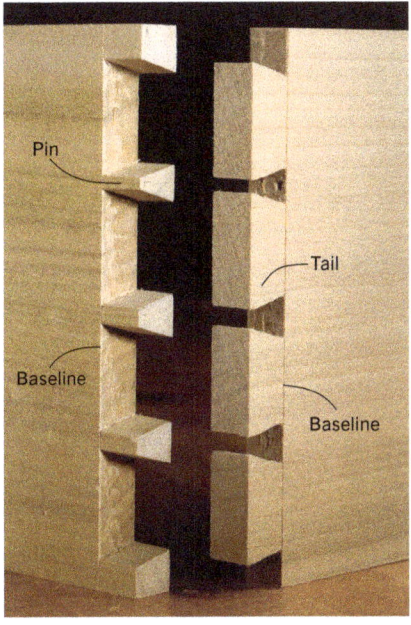

Dovetail anatomy. The essential parts of a through-dovetail joint.

HAND-CUT DOVETAILS | Your First Hand-Cut Dovetails

Lonnie Bird's essential toolkit for dovetailing. (From top): A mallet, a marking gauge, a chisel (note the shortened handle), an adjustable square, a dovetail saw, a dovetail marker, more chisels, and a knife.

Too square (far right). Many modern chisels have sides that are too square for getting into the triangular sockets between the joint's tails (left). I recommend grinding the sides down almost to the back (right).

All dovetails have baselines; the baseline indicates the height of the tail or pin.

TOOLS

The tools for dovetailing are not expensive but it's important to have the right ones. It's also important to have them well-tuned.

Before cutting a dovetail you'll need to do a bit of measuring and marking, commonly referred to as layout. Good layout is essential. Remember: Dovetailing is the simple act of sawing and chiseling to a line; if the line is inaccurate, the joint won't fit together.

One of the most important tasks is marking baselines. The baseline is created with a marking gauge—a simple tool that consists of a head, beam, and cutter. The head slides along the beam and locks in place with a thumbscrew. Some gauges use a steel pin for the cutter while others use a tiny wheel. Either type of cutter will work so long as it's sharp. A dull marking gauge will tear the fibers, making it difficult or impossible to craft a clean dovetail joint. In contrast, a sharp gauge will cleanly sever the tough end-grain fibers to create an incised layout line. As you chisel out the waste between the tails and pins, the edge of the chisel will drop precisely into the baseline to give you that great fit that you're striving for.

Other layout tools you'll need include a layout knife, a square, and a dovetail marker. A craft knife works well; it's razor sharp and the narrow point will easily scribe between the tails and pins. The type of square is unimportant as long as it is 90°; I prefer combination squares for their precision and versatility. To mark the angle of the tails and pins, I use a dovetail marker. One with a simple extruded aluminum design is best; they are inexpensive and I can rework the soft aluminum to an angle of my choosing, typically 14°. A 14° pitch provides the good looks and mechanical interlock that I'm always after.

Of course, you'll also need a dovetail saw, a few chisels, and a mallet. There are two types of dovetail saws available today: Western and Japanese. Traditional Western-style dovetail saws cut on the push stroke and feature a thick back to stiffen the blade and prevent it from buckling. However, Japanese saws cut on the pull stroke, which places the blade in tension during the cut so it doesn't have the tendency to buckle. Consequently, Japanese saws have a thinner blade and cut a finer kerf. Also, the unique tooth design of the Japanese saws cause them to cut more aggressively than Western saws. Which is best? When I teach dovetailing I encourage people to experiment with each. Although most choose the Japanese saws, others feel they get more control and a truer cut with the Western saw. Regardless of which style you prefer, it's important to use a high-quality dovetailing saw.

The best chisel for chopping waste from between the tail and pins is a short one. A short chisel provides the control you need when driving the chisel with a mallet; long chisels are designed for paring. For many years I've used the long-discontinued Stanley #750 socket chisels. The short 9" length and perfect balance of these old tools are just what's needed for dovetailing. Stanley #750s are still available from old tool dealers, and Lie-Nielsen Toolworks has begun manufacturing its own improved version of these venerable chisels. Of course, if you already own a set of inexpensive chisels, you can also do what many of my students do—cut the excess length from the handle. Although it may sound odd, reducing the handle length greatly improves the balance of a long, top-heavy chisel. The improvement will be reflected in the quality of your dovetails.

An important step to fine-tune chisels for dovetailing is to further bevel the sides of the blade up by the cutting edge. On most new chisels, the sides are too square, and the excess steel crushes the fibers of the tails and

Mark the baseline. Do so on the faces of both the pin board and the tail board with a marking gauge. Also mark the baseline on the long edges of the tail board.

Mark the pin shapes. Use your dovetail square to mark the shape of the pins on the end grain. Then mark the face of the pins using your adjustable square.

HAND-CUT DOVETAILS | Your First Hand-Cut Dovetails

Start the saw. Use your thumb as a guide to start the kerf of your saw in the edge closest to you. After a couple strokes, begin to lower the angle of the blade.

Extra kerfs. To make the waste between your pins easier to remove, cut several extra kerfs in the waste. Take care not to cross the baseline of the joint.

pins as you chisel the waste. Grinding the sides close to a knife-edge will eliminate the problem (see p. 12, top right). Of course, you should also hone the chisels to razor sharpness.

Having the right mallet is important, too. I've found that a round, 12-ounce mallet works best. Heavier mallets are tiring to use and the extra weight just isn't needed. Also, the head of a square mallet must always be aligned to the chisel before striking. Not so with a round mallet.

Once you've gathered your tools and tuned them up, you're ready to begin.

LAYOUT

The first step in the layout process is to mark the baselines. Note that the baseline is marked on both faces of both halves of the joint. It's also necessary to mark the baseline on the edges of the tail board. First, set the gauge to the thickness of the stock. As you mark the baseline, focus on keeping the head of the marking gauge firmly against the end of the stock. To avoid tearing the grain, make several light passes with the gauge as opposed to one heavy cut.

Mark the half pins on each corner of the pin board, and then divide the board into the number of desired tails. Each point of the divider becomes the center of a pin. After marking the slope of the pins on the end of the stock, mark the face with a square.

SAWING

As I stated earlier, dovetailing is essentially sawing and chiseling to a line. Once you've mastered that technique you can cut great-looking dovetails.

Start by positioning the saw on the near corner of the stock and pull the saw to establish a small kerf. During this initial cut, it's helpful to use your thumb to guide the saw. As you pull the saw toward you, lower the blade into the stock to establish the top line.

Now use long, smooth strokes to follow the line on the face of the stock. Stop when you've reached the baseline. Once you've sawn all the pins, make several extra saw kerfs into the waste area between the pins. These cuts will make it a lot easier to chisel the waste between the pins.

Next, select a narrow chisel, ⅜" or ½", and make certain that it is razor sharp. A narrow chisel has less cutting resistance than a wider chisel and you'll have better control of the tool.

To remove the waste between the pins, it's best to cut halfway through the stock from each face. But remove the bulk of the wood first by positioning the chisel about 1/16" away from the baseline. Drive the chisel halfway through the stock, flip the stock over, and repeat (see sequence photos on p. 15).

Position the edge of the chisel in the baseline (note how easily it drops into the incised line) and repeat the process. It's good practice to undercut the baseline very slightly. The undercut surface ensures a tight fit and doesn't weaken the joint. (Remember that the strength comes from the interlocking tails and pins as well as the long-grain gluing surfaces.)

Now examine the end-grain surface very closely. You should see a fine line along the edge of the stock that was created by the marking gauge. If you don't see this line, you've chiseled too far—or not far enough.

LAY OUT THE TAILS

The tail board layout is created from the pin board. First, position the tail board face down on the bench. Next, place the pin board over the tail

Remove waste between the pins. First, position the chisel 1/16" from the baseline and cut halfway through the thickness of the board.

Flip. Flip the board over and do the same on the opposite side.

Angle the chisel. To chisel out the rest of the waste, place your chisel into the baseline and undercut the joint just a bit by angling the chisel as shown.

■ HAND-CUT DOVETAILS | Your First Hand-Cut Dovetails

Halfway on each side. Cut halfway through the waste on one side. Flip the board over and repeat.

Baseline remains. Here you can see what's left of the baseline after removing the waste. This fine line is the evidence that you've chiseled to the correct point.

board, align the face with the baseline of the tail board, and then clamp it in place. Remember, too, that the wide part of each pin should be facing the inside of the joint.

Mark the tails with your layout knife. Position the blade of the knife against the pin and then use the pin to guide the cut. To complete the layout, mark the end of each tail with a knife and square.

Sawing the tails is similar to sawing the pins, except you'll have to tilt the blade on the vertical axis. I think it's bad practice to angle the tail board in the vise; it's best to learn to angle the saw instead. Otherwise, when sawing the tails of a wide board for large casework, one corner of the board will be positioned high up in the air, which will make sawing difficult. Instead, clamp the tail board in the vise (make sure it's level) and saw all the cuts one direction. Then, saw all the cuts that are angled the opposite direction.

I use the same technique for chiseling the waste as I use on the pins; make a few extra saw kerfs and chisel halfway from each face. Remember to undercut this end-grain surface slightly. However, be careful to not undercut the surfaces at each corner. Otherwise, you'll see a distracting void in the assembled joint.

To assemble the joint, first position the pin board upright in the vise. Gently press the tail board into the pin board using pressure from your thumbs. When assembling dovetails on wide casework, I'll use gentle taps from a dead-blow mallet. You can hear and feel where a portion of the joint may be too tight. Simply pare a shaving from any such areas, slide the joint together, and step back and admire your work. With patience, you'll find that dovetailing is one of woodworking's most pleasurable tasks. ■

Your First Hand-Cut Dovetails | HAND-CUT DOVETAILS

Mark tails. Clamp your pin board to the mating tail board and transfer the shape of the joint to the tail board using your marking knife.

Clamp it up. Clamp your tail board vertically in your vise and saw the shape of the tails. Make a few extra kerfs in the waste and chisel it out much like you did with the pins.

Complete chiseling. When you remove the waste between the tails, slightly undercut the end grain between the tails—except on the ends, where it will show.

Assemble. In wide casework especially, you may need a few taps of a dead-blow mallet.

HAND-CUT DOVETAILS | Your First Hand-Cut Dovetails

A DOVETAIL A DAY

BY CHRISTOPHER SCHWARZ

I was on our local swim team as a child, and I was an embarrassment to my pool, my parents, and mammals in general. Perhaps the coach kept me around to make the youngest swimmers (Team Minnow) feel better about their dog-paddling. Or perhaps my artless splashing lulled competing teams into complacency before a swim meet.

One summer day my mother dropped me off at the pool, and as she drove off I discovered that none of my friends were there. I had the entire day alone before me.

I got in the pool and messed around a bit. As boredom set in I swam a couple laps of breaststroke. After a few laps

I wondered if I could stretch my hands forward more. I then wondered if I could tuck my legs in tighter after a kick. Three hours later my mom called me from the pool side to go home.

The next day was a swim meet, and I was in the 50-yard breaststroke against kids who beat me every summer. The starting gun fired, and 50 yards later I looked around. I was alone. I had won by an enormous margin. It was my first and last victory.

You know where this story is going.

Now I've always been a fair dovetailer. I cut my first set by hand years ago and made decent joints. But I was slow. One day the memory of that swim meet returned, and I decided to try the same approach with my dovetailing. I vowed to cut a dovetail every day for a month.

That night I prepped a few boards of cherry and poplar. I laid out my tools on the bench and cut my first set—three tails into three pins. It took more than an hour. I then cut the joint free of the two boards, marked the date on the corner and put the joint on the windowsill. I left all my tools out on the bench, set and ready for day two.

The next day, before I cut the second set, I picked up the joint from the night before. Under scrutiny, it wasn't as nice as I'd remembered. My saw had crossed the baseline here. I had split one pin slightly there.

I cut my next set and tried to avoid crossing the joint's baseline. I tried to ensure the pins on the ends were cut straight. And I made the half pins on the ends a bit wider.

I cut that joint free, dated it and sat it on the sill. After a few more nights I realized that I was just repeating my blunders. Split pins were plaguing me.

So I sawed even closer to my knife lines on the end pins. The next day, no splits.

After two weeks, my dovetails looked tighter. Then I changed their spacing. Then I started to pick up speed and arrange my tools so I wasn't fumbling for the chisel.

After 30 days, I was 10 times the dovetailer I was when I began. The operation felt natural. When the 30 days was up, I was worried about stopping my experiment. Would I regress? That had happened when I was on the swim team. I had stopped swimming practice laps and never won another race.

But this story has a happy ending. Once I conquered the dovetail, I used the joint more often in my work. I also began sawing and chiseling more in general, which then reinforced my dovetailing.

So many times we learn woodworking on the fly as we build something. We get our skills just good enough to accomplish that project and then we move on. It's rare to get out a board and just saw it. Or plane it. Or mortise it with our router.

This method might seem like wasting time, but it has resulted in some of my most enjoyable shop time. And now I'm thinking that "Inlay a Day" has a nice ring to it.

4 WAYS TO MAKE DOVETAILING EASIER

While cutting a dovetail a day, here are the four small breakthroughs I had that have helped me with hand-cutting dovetails over the years.

1 Rip teeth. I used to use a crosscut dozuki to cut dovetails. It worked fine, but rip teeth track better. Saw teeth on Western saws are more robust (I have ruined several Japanese crosscut saws during what I would consider normal use). I'd purchase either a rip Western saw or a rip dozuki. Rip teeth do make a difference.

2 Corner clamps. One of my biggest mistakes historically was not correctly transferring my tail layout to my pin board. To mark the pins, clamp the tail board in place with a set of inexpensive 90° corner clamps, available from any home-center store. This clamping setup allows you to focus on correct marking.

3 Switch to a marking knife. I was using a pencil to lay out everything. My joints got better when I started using a spear-point marking knife. Fine lines make a difference.

4 Learn to darken the lines. Knife lines are hard to see in some woods. After knifing my lines, I run over them with a mechanical pencil (the version with a .5 mm lead). Its mark is pretty coarse compared to a knife line, so I quickly and lightly run a quality eraser over the line. A fine knife line that's filled perfectly with dark lead is left. It's easy to see and track. I know it sounds like an elaborate procedure, but it takes just a moment (see illustrations below).

4. LEARN TO DARKEN THE LINES

A. Knife. Mark your lines with a spear-point marking knife.

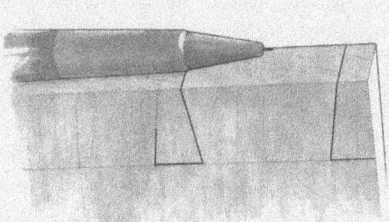

B. Pencil. Darken the knife lines with a mechanical pencil.

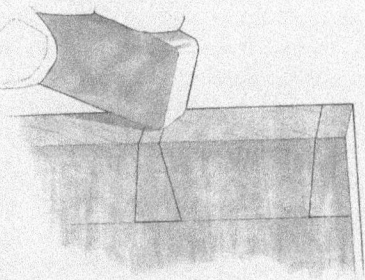

C. Eraser. Erase the excess lead. This leaves very sharp and dark lines.

FRANK KLAUSZ ON LEARNING TO CUT DOVETAILS

Stop measuring and simply learn how to saw straight.

BY FRANK KLAUSZ

The dovetail is an ancient joint widely used in cathedrals, barns, and Egyptian furniture. It is the right joint for many items, including fine furniture, carcases, drawers, and jewelry boxes. They are all dovetailed together.

I was only 27 years old when I came to this country in 1968 from my native Hungary. Although I had a piece of paper that said "master cabinetmaker," I was still very eager to learn more about my trade.

Where I came from, I was happy if I could carry a white-haired master's tool chest to the job site because I knew I would learn a thing or two that day working with him. Now I am that white-haired master with 45 years of experience in the trade.

In the early 1970s, I went to a lot of seminars. Some were on dovetailing with well-known teachers in the woodworking world. Some cut the tails first; others cut the pins first. They used tools that I didn't own, such as a dovetail marker. They measured the size of the pins and tails, which is completely different from my method. The more I studied, the more confused I became. I decided to

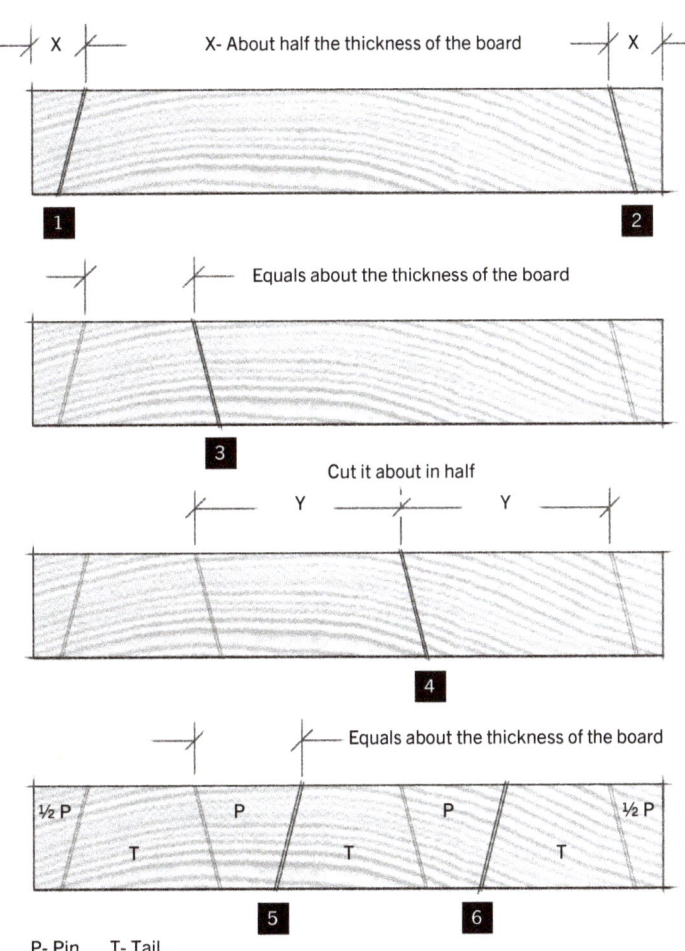

THE DOVETAIL BOOK | 21

HAND-CUT DOVETAILS | Frank Klausz on Learning to Cut Dovetails

find the best way to cut tight dovetails quickly.

A SEARCH FOR THE BEST METHOD

I owned an antique restoration shop. I had a chance to study a lot of antiques from around the world. Each time a piece of furniture came to the shop, the first thing I looked at was the dovetails. I studied hundreds of them and made tracings of dozens of unusual pieces. I tried to find an answer for my methods. I learned in Hungary, I worked in Vienna, and I was looking for someone from a different part of the world than Eastern Europe to do dovetails. I found Hector, from Guatemala, a master cabinetmaker.

"This is great, Central America!" I said. I asked him to make me dovetails. He said, "You cabinetmaker, you make dovetail." We had a language problem. I had a hard time explaining to him my intentions. I replied, "I know how to cut dovetails, I want to see how you do it." "OK," he said. He grabbed some chisels, a dovetail saw, a marking gauge, some scrap wood, set up the marking gauge to the thickness of the wood, marked the wood, clamped it into a vise and started cutting. He cut the pins, then chiseled the pins; he marked the tails from the pins, chiseled the tails, and put it together. "How is that?" he asked. I was as happy as could be! "That is exactly the way I do it," I replied.

After my experience with Hector, and my 10 years of researching dovetail techniques, I came to the conclusion that Grandpa wasn't a bad craftsman at all and my father taught me well.

Later on, I wrote some articles for different magazines and I made some instructional videos, some about dovetails. Before I knew it, I was teaching the craft throughout America. I taught hundreds of people how to dovetail. A lesson took plus or minus one hour with a 99 percent student success rate (let's face it, some of us are born with two left hands).

ANYONE CAN DO IT

If you already know how to do dovetails, and are happy with your method, I am happy for you and don't mean to change your ways. If you are a beginner or learning about new methods, you can do it my way. I know you can do it!

How do you know how to write? You learned in school. You made a whole row of *a*'s. You made a whole row of *b*'s. Before you knew it, you were writing words and sentences. That's how I learned to do dovetails. In school, I cut a whole row of straight cuts without marking, checked it often with a square, and improved the next row. In the next lesson, I cut angles approximately 10° to 15°, all to the left, the next row all to the right, and before I knew it, I was cutting dovetails.

CUT DOVETAILS EASILY ON BIGGER BOARDS

When cutting dovetails on a wider board, use the same method as I describe here. You have to divide the remaining space after your third cut in half and half again, or ⅓. With practice it will come naturally. The thicker the wood, the bigger the pins and tails. For example, a 1"-thick board for a blanket chest should have 1" to 1½" tails. It both looks good and is very strong. When I was an apprentice watching my father work, I asked him, "How can you do this so fast?" He replied, "Don't worry, after 10 to 15 years you will be a good beginner yourself."

Frank Klausz on Learning to Cut Dovetails | **HAND-CUT DOVETAILS**

Find thickness. Set up the marking gauge exactly to the thickness of the wood.

Mark the wood.

Cut a half pin.

Cut another half pin.

Cut a tail.

Find the next one. Divide the distance in half between the two saw kerfs and cut it.

Companies sell router bits from 7° to 18°, so the angle you use is a personal choice. The strongest dovetails have equal-sized pins and tails, like machine-made drawers. Pope John Paul II's coffin had approximately 3" pins and 3" tails (see p. 27). The choices are endless.

CUTTING DOVETAILS MY WAY

So how do you make dovetails my way? Make yourself a cheat sheet (see the drawing on p. 21) or look at some dovetails to copy. Get some scrap wood. Mill them to the same size: 3½" to 4" wide, ½" thick, and 5" to 6" long. Mill five, 10 pieces; whatever

HAND-CUT DOVETAILS | Frank Klausz on Learning to Cut Dovetails

Go back to your first angle. Cut another pin.

Cut one more pin. You're done cutting pins.

Ready to chisel. Put the chisel into the marking gauge line and tap it.

Repeat. Do the same on all the tails.

Time to tilt. Tilt the chisel forward to take out a little piece.

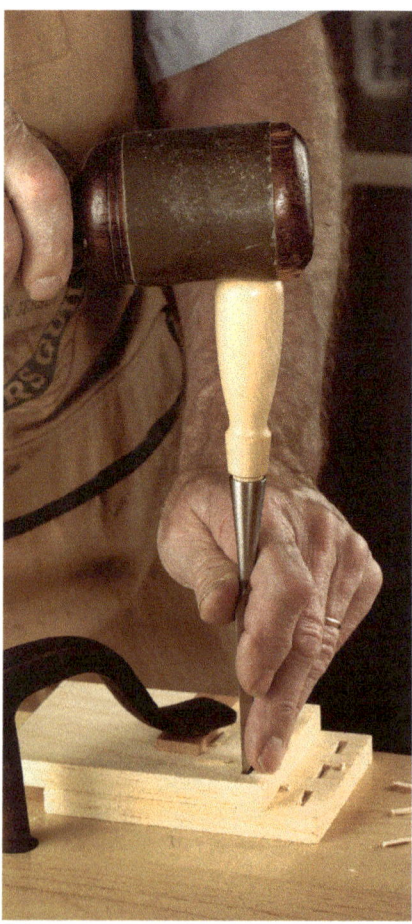

Straighten. Put the chisel back up and tap it more. Undercut a little bit, approximately 2°.

it takes. Set up your marking gauge exactly to the thickness of the wood. Mark the face of the wood. Clamp it into your bench vise, and start cutting with your dovetail saw. (I hope you already practiced your rows of straight and angled cuts.)

Every dovetail starts with a half pin. On the other side is another half pin. Cut them. Next to the half pin you need a full tail. Cut it. Cut the remaining distance in half with the same angle, turn it around, make two more cuts, and you're done. Cut only pins, and cut as many as you need until you are pleased.

There's no marking involved—use your eyesight and judgment, and use the thickness of the wood for the width of the tails by judging distances. Make them to your liking. My pins are a little smaller than the tails. That's the way I like them. You may make them the same way or you may

24 | THE DOVETAIL BOOK

make small pins such as ones found in English furniture. They are all good. You are cutting hand-cut dovetails; there should be some variation. Hand-cut dovetails have character and Mrs. Jones likes that.

Once you're happy with your pins, chisel the pins. Put the chisel on the marking gauge line and tap it. Take out a little V cut. Now chisel deeper, taking out chips. Undercut just a very little. Flip the piece over and do the same on the other side.

Next, use your pins to make the tails. Hold all three sides even with the edge and the end. With a sharp pencil, mark it from the inside. Here is the hard part: When you cut the pins a little this way or a little that way, it doesn't matter because you're making a template. But when you cut the tails, you have to be accurate and cut that pencil line in half. To understand which side of the pencil line you are cutting, you have to mark the half pins and pins with an X. That will be your waste. When you chisel out your waste, the X will become sawdust and chips. Cut off your half pins; chisel your tails (you are chiseling out the space for the pins).

Here comes the fun part: Try fitting it together. If it doesn't fit, try to find out why, but don't fix it. Cut your next piece. You may have to go closer to the line if it is too tight or leave more of the line on to make it tighter. Make a new one using the same pins until you are happy with a snug fit.

You are ready to make a jewelry box for your mother-in-law. Good luck trying; I am sure you can do it! Happy woodworking. ■

Corners. With a smaller chisel, chisel into the corners. Chisel about halfway.

The other side. Flip your stock and chisel from the other side.

Mark the tails. Use the pins to mark the tails. Hold the pin board flush on the outside and on the edges.

Mark the waste. Mark the bits of wood you will be cutting out with an X.

HAND-CUT DOVETAILS | Frank Klausz on Learning to Cut Dovetails

First half tail. Line up your saw with the pencil line. Use your thumb for a guide and cut on the *X* side.

Check your lines. Here you can see what it looks like to leave the lines on the tails.

Cut off the half pin. The saw kerf should be outside of the marking gauge line.

Chisel the tails. Proceed the same way you chiseled the pins.

Remove the waste. With the edge of your chisel, push out your waste.

Put them together. If you did everything right, it should easily tap together.

Complete. Here you can see the finished practice piece.

A DOVETAIL STORY: THE POPE'S COFFIN AND THE UNIDENTIFIED CRAFTSMAN

BY KARA GEBHART UHL

The morning after Pope John Paul II's funeral, John Darrow, Frank Klausz's finisher, asked Klausz, "Did you see the pope's coffin?" Klausz hadn't. "It has big, big pins and tails, just like you do them!" Darrow said.

Using pictures of the coffin to determine scale, Klausz made a replica pine corner. Then he examined the joinery in the pictures using a magnifying loupe. "It was easy to tell nobody measured or used angle gauges," Klausz said.

Finding out who built the highly publicized coffin should have been easy, but it wasn't.

The quest first led to the editor of the Catholic magazine *America*. He said to contact the United States Conference of Catholic Bishops, who said to contact the Vatican Embassy in Washington D.C., who said to contact the Vatican press office, who said they don't handle requests "for such small details." Several other publications and an organizer of a funeral fair in Poland didn't respond.

Several Vatican experts did research for us and found nothing. The Catholic News Service also couldn't provide detailed information.

Two experts who were in Rome asked several people in the city but no one knew anything. Rev. Steven M. Avella said Archbishop Stanislaw Dziwisz, "the pope's dear friend and personal secretary," knew but doesn't respond to these types of requests.

Wendy J. Reardon, author of *The Deaths of the Popes* and part-time teacher of exotic dancing (yes, exotic dancing) suggested contacting Alan Howard, who runs St. Peter's Basilica's web site. "He's got some powerful friends in St. Peter's, so perhaps he can ask them," Reardon said. Howard gave us the address of Archbishop Piero Marini, who planned the pope's funeral. John-Peter Pham, author of *Heirs of the Fisherman: Behind the Scenes of Papal Death and Succession,* also suggested we contact Marini. So we sent a letter to Vatican City.

Surprisingly, Marini sent a letter back. After translating its few sentences we learned this: The coffin was built in Vatican City.

While this news was welcome, the lack of further information was unfortunate. The quest took some alternate paths. Travel writer Rick Steve's consulting department was unable to provide information about Vatican woodshops. Brian Boggs, a chairmaker in Berea, Kentucky, provided contact information for Thom Price, a gondola maker in Venice, who didn't respond.

Pallbearers carry Pope John Paul II's dovetailed cypress coffin into the Basilica during his funeral in the center of St. Peter's Square at the Vatican on April 8, 2005. Note the coffin's large pins and tails.

But Mark Marsay, a London, England–based refinisher and tool restorer with family in Italy, recalled hearing that the coffin was made by Vatican Museums' restorers and conservators. He also didn't think it was a solo effort. But he couldn't confirm anything.

Our quest ended when we called Vatican Museums: Our deadline had approached and no one on the other line spoke English.

Although much remains unknown, we did uncover some interesting facts and suspicions:

■ The cypress coffin was adorned with a carving of a cross and the letter *M*, which stood for "Mary, the mother of Jesus."

■ Along with the pope's body, a sealed lead tube called a *rogito* containing a sack of bronze and silver Vatican medals and a brief biography written in Latin were placed in the coffin.

■ The pope was buried in three coffins. The innermost cypress coffin was placed inside a zinc coffin, which was placed inside another wood coffin (some sources say the wood is elm; others say it's walnut or oak).

■ Charles Garnette, an Indiana-based woodworker who plans to build a replica of the cypress coffin, surmised from pictures that the wood is about 1" thick. Garnette says there is speculation by Vatican historians that the cypress was recycled wood, perhaps old door planks. He also believes the coffin actually was built in 1981, when Pope John Paul was shot.

Still, it's too bad we don't know who built the coffin. Such skill and dovetail methodology deserve recognition and exploration.

DOVETAIL RULER TRICK

A throwaway wooden ruler prevents fatal errors when dovetailing.

BY CHRISTOPHER SCHWARZ

The rabbet trick. For years I've shown students how to cut this shallow rabbet to make it easy to transfer the shape of one board to another during dovetailing.

The number one mistake made by first-time dovetailers has nothing to do with sawing or chopping—the obvious choices.

Instead, I've found that most fatal mistakes happen at the point where the shape of the first half of the joint—the tail board or pin board—is transferred to its mate.

During the transfer process, beginners fail to align the boards properly, or a board shifts during the transfer process. The end result is that the joint is horribly misaligned or, worse, it won't go together.

To fight this alignment problem, I used to show beginners how to cut a shallow rabbet on the inside of the tail board to help the two boards mate easily during the transfer process, reducing errors.

This strategy works great—if you can cut a square, well-placed rabbet. To be honest, it is difficult to teach beginners to do this with a rabbet plane at the same time they are also learning to knife, chisel, and saw a dovetail joint.

I was beginning to wonder if the rabbet was more trouble than it was worth.

JUNK DRAWER INSPIRATION

One day I was pawing through a bin of tools and parts that I'd been meaning to get rid of when I came across a stack of wooden 12" rulers branded with advertising—the kind you can often get free at hardware stores.

Something clicked. I grabbed the rulers and strode to the shop to experiment with some dovetails with this idea in my head: Instead of cutting a rabbet to help register the pin board and tail board, could I create a "rabbet" by tacking a ruler to the baseline of the tail board?

The answer is "yes," and it has made teaching dovetailing a great deal easier for me.

ABOUT THE RULER

All you need to try this is a wooden 12" ruler (or a paint-stirring stick) and a couple of nails with narrow shafts and sharp points. Escutcheon pins are a good choice, as are headless brads.

The goal is to fasten the ruler with one of its edges on the baseline of the tail board. Pins work best for me, though you might consider using double-sided tape or hot-melt glue.

Before tacking the ruler down, I drive the nails most of the way through it. Then I place the ruler on the baseline and drive the pins into the tail board about 1/8" or 3/16" deep. (If you have trouble aligning the ruler on the baseline, try driving in one fastener only, make any final adjustments then drive the second.)

That's the trick. Here's how it works.

Big or small. It matters not if the ruler is shorter or longer than the tail board's width. All that matters is that it's planted on the baseline.

On your mark. The ruler allows the tail board to lock to its mate right at the baseline. Now you can focus on transferring your layout.

HAND-CUT DOVETAILS | Dovetail Ruler Trick

Yup, it works. If you cut your dovetails pins-first, you can still use a ruler to help align the two boards during the process.

Don't miss your mark. Because of the thickness of the ruler, you won't be able to mark all the way to the baseline when cutting half-blind dovetails pins first (left). But the slope is already set by the saw's kerf by the time you reach this area (right). So it's not a problem.

MAKE THE TRANSFER

Now transfer the shape of one board to the other. Place the tail board with the ruler against the pin board, which I place upright in a vise. Shift the tail board to align the two at their long edges.

Use a knife or pencil to scribe the shape of the joint onto its mate. If you cut your tails first, this trick works easily with both through-dovetails and half-blind dovetails.

FOR PINS-FIRST DOVETAILERS

This trick also works if you cut your dovetails pins first. Cut your pins as usual. Affix the ruler to the baseline of the tail board. Then place the pin board against the ruler and draw in the shapes of the tails on the tail board.

The only fly in the ointment is when cutting half-blind dovetails pins first. Depending on the thickness of your ruler, you might not be able to mark all the way down to the baseline of the tail board. My ruler was a bit thicker than 1/8" and I could mark down to almost 3/16" from the baseline. This isn't a big deal—the shape of the tail is already set by the saw's kerf by the time you reach the base of the tail.

After the transfer is complete, pry the ruler off the tail board and use it again. My favorite part of this technique is wondering what the furniture conservators of the future will make of the two little holes on my tail boards. ■

DOVETAIL DILEMMA

What's more important: Strength or aesthetics?

BY GEORGE R. WALKER

For me, a high point of the Woodworking in America conference over the years was the dueling dovetail session between Roy Underhill and Frank Klausz. The two squared off with saw and chisel in hand tackling the "pins first vs. tails first" debate. Friendly banter peppered the dialogue as these two masters cut dovetails with an ease and deliberateness that spoke volumes. Both represented a woodworking tradition, with Frank "Pins First" Klausz demonstrating skills learned in an Eastern European woodshop, while Roy "Tails First" Underhill shared his wisdom of historical American craft. But one part of the discussion in particular caught my attention.

Frank pointed out that his Grandpa taught him to make the joint strong by sizing the pins and tails equally. He even had a photo of Pope John Paul II's funeral casket, a simple box knitted together with bold, equally sized pins and tails (see p. 27).

It begs a question: Why would American and English cabinetmakers choose a different path and intentionally break up the sizing and spacing on dovetails? They typically make the tails large and the pins small, sometimes delicate. This can sacrifice strength in certain applications and can make the joint more

THE DOVETAIL BOOK | 31

HAND-CUT DOVETAILS | Dovetail Dilemma

Large and small. This chest of drawers exhibits typical dovetails from American work with large tails and small pins.

difficult to execute. It's hard to say for certain, but my gut tells me they did this for aesthetic reasons. This was a craft tradition so familiar with proportions that even half-blind dovetails hidden away in a drawer should give a nod to beauty.

This also leads to questions about joinery and design. Aside from the obvious structural role that joints play, can we use them to enhance a design aesthetically? Should joinery be hidden or left in plain sight? If we choose the latter, how can we make sure it complements our design?

JOINERY AS STRUCTURE

Any time we join materials together, whether building a house or a cabinet, we are creating structure. A question facing a designer is whether to leave structure exposed or hide it from view. There is no right or wrong answer. Cabinetmakers during the Federal era (circa 1790–1810) took great pains to hide any traces of joinery. A century later, Arts & Crafts furniture makers took an opposite tack and purposely exposed much of the joinery. Makers in both eras made sturdy, beautiful furniture with very different ideas about whether joinery should be hidden or in plain sight. So, when does it make sense to hide joinery or leave it in the open? It might help to look at architecture for some answers.

Designers from antiquity often covered structural elements such as brickwork, yet took great pains to include stylized representations of ancient joinery. Much of the familiar ornament on classical work shadows the original joinery on ancient structures built with wood. The triglyphs that adorn the Doric entryway below left represent the carved wooden beams that would have supported a building. Dentil moldings are another example of a stylized depiction of structure hinting at ancient joinery supporting a roof. Although the triglyphs are purely ornamental on this doorway, they succeed in highlighting the opening. They proclaim, "Here's the front door; welcome!"

JOINERY HIGHLIGHTS A FORM

The subhead above is an important point. By its very nature, joinery holds together major elements. As is often the case, those elements also define the boundaries of the form.

Bold. Triglyphs over this Doric doorway make a bold visual statement.

Contrast. Dentil molding creates a contrasting light and shadow line, outlining the form from a distance.

Dovetail Dilemma | **HAND-CUT DOVETAILS**

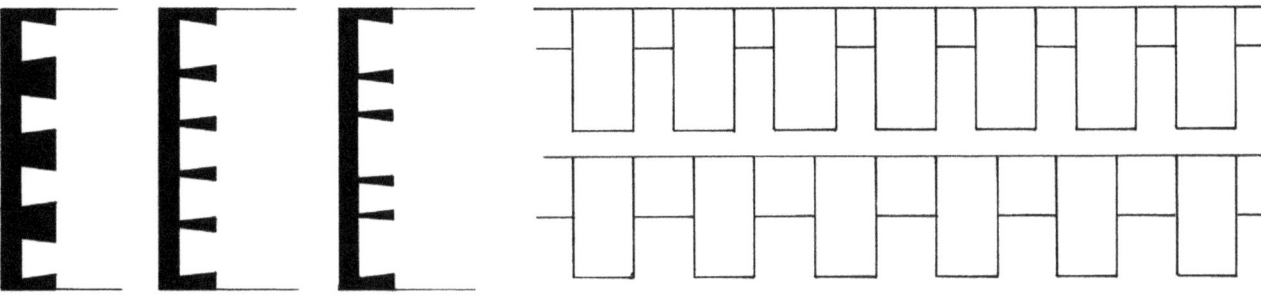

Arresting. Which dovetail sequence catches your eye?

Preference. The dentil sequence on top alternates major and minor spacing 3:2:3:2:3. The bottom sequence repeats 1:1:1:1. Which one does your eye prefer?

A form is a combination of simple shapes that anchors a design, often made up of rectangles, circles, or ovals.

Designers often use moldings, carvings, or inlay to highlight a form and give the eye something to grasp from a distance. Joinery such as dovetails can also be used as a border element to highlight a form or a sub-element within a form. Tenons or pegs can also be left exposed at the corners or borders to gently emphasize a form. The key is to use them as a highlighter.

When we highlight text with a yellow marker in a book, we make the words stand out. If the highlight is too bold, it might obscure the letters, defeating our purpose. If we make the joinery (or moldings for that matter) too bold, they compete with rather than emphasize the form. Joinery should do more than just show off some fancy chisel work. Rather, exposed pegs or joints can subtly complement the underlying bones of a form.

DOVETAIL DEBATE
Confession time: I cut dovetails "tails first." It took quite a bit of practice to master the technique and achieve tight joints that I'm pleased with. But, after watching Frank, I may have to give "pins first" a whirl. He made it look almost effortless, and with stunning results. I also agree with Frank that pins and tails sized equally produce a strong joint. Yet like Roy, I'll sacrifice a little strength for aesthetics. Sizing the opposing pins and tails into major and minor spacing gives it some visual flavor.

Look at the dovetail patterns shown top left; which appeals to your eye? Note the dovetail pattern on the far right is made livelier by adding another layer of major and minor spacing. Take another look at the proportional sizing and spacing on a simple dentil molding at the top right of this page. The face of a dentil block is twice as high as the width. The spacing between blocks is two-thirds the width of the block. This puts the blocks in a major-minor sequence. Compare that with a series of blocks sized and spaced equally.

Pins vs. tails? One thing's certain; in 200 years, a well-executed dovetail will still show the skill of the artisan's hand. ∎

THE DOVETAIL BOOK | 33

4 TIPS FOR DOVETAILING BY HAND

Improve your technique with these pointers.

BY FRANK KLAUSZ

Follow along with these tips to make dovetailing by hand go more smoothly.

1. CUT BY EYE

When dovetailing by hand, be bold. Don't bother with a try square or sliding bevel. After gauging a thickness line across the board, lay out the pins with your saw as you make each cut. Trust your eye to find a pleasing dovetail angle and repeat it over and over. Trust your hands with a sharp saw to cut straight down. If you've never done this before, let go of your anxiety and just do it. Practice, practice, practice. You'll be amazed at how easy it becomes.

Don't worry about the exact sizes of the pins or the spaces between them. In fact, a little variation is a good thing. Your right and left hands aren't perfectly symmetrical, so it's okay if your dovetails aren't, either.

Follow a "rule of halves" when you saw the pins. First, cut two half pins, one on each end. Then cut the other side of one tail.

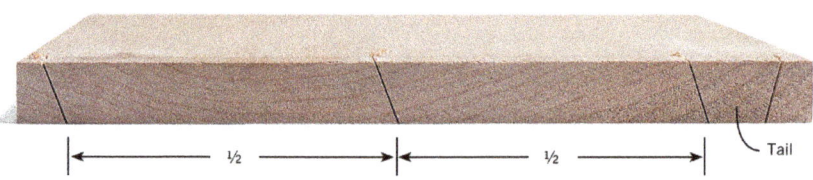

Eyeball half the distance between the tail and the opposite half pin and make another cut.

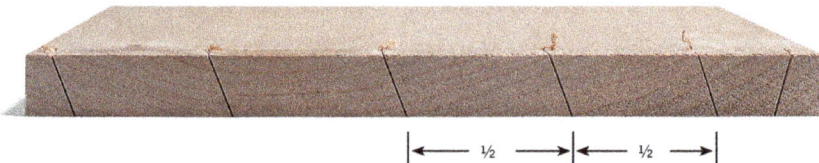

Divide the remaining spaces in half again with additional cuts.

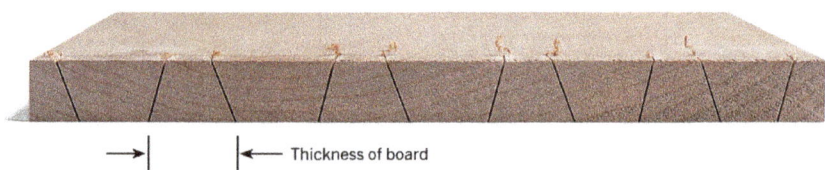

Turn the saw in the other direction and saw each complete pin. Judge their width by looking at the thickness of the board.

4 Tips for Dovetailing by Hand | HAND-CUT DOVETAILS

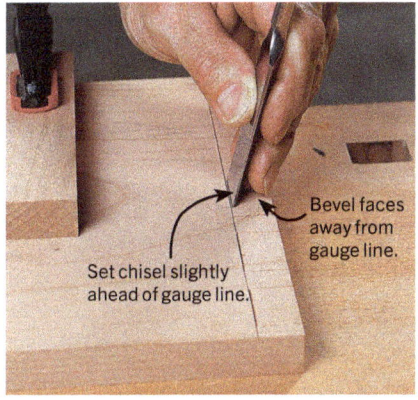
Set chisel slightly ahead of gauge line.
Bevel faces away from gauge line.

Flat spot
Tilt chisel away from gauge line.

2. CHOP WITH VIGOR

To start, plant the chisel slightly ahead of the gauge line and give it a whack with a round-headed mallet. The slope of the chisel's bevel moves the cut right to the line. Experience tells you how far away from the line to set the chisel. March across the board, repeating the same cut.

Next comes a sloped cut that defines the shoulder. It's made with the chisel facing the same direction as the initial cut—an economy of movement that's common to all good handwork. This cut carries the chisel pretty deep, almost up to the first cut.

A combination of vertical cuts and sloped cuts remove most of the waste. Slightly tilt the vertical cuts. This "undercutting" saves a lot of time in cleaning up the dovetails.

Leave a flat spot at the end of the waste. If you've ever torn out the interior of a hand-chopped dovetail, you'll appreciate this tip. Chop halfway down one side of the board, brush away the chips, and turn the board over. The flat spot supports the waste so there's no tear out.

Wedge fills gap in joint.

3. FIX MISTAKES

Don't worry about making a few mistakes—just learn to fix them. Here a mistake in cutting left an open joint. So while the glue was still wet, a wedge was sawed and lightly tapped into the gap. Now that the glue has dried, the end of the wedge can be off.

4. LEVEL QUICKLY

Use a plane to clean up, working from the corners to the middle. When finely set and super sharp, a handplane slices through the end grain, leaving a smooth, flat surface that you can't get with sandpaper. Remove any ridges the plane may leave with a cabinet scraper. ■

PRECISE HAND-CUT DOVETAILS

This new approach to half-blind dovetails guarantees a precision fit.

BY TOM CASPAR

The handmade half-blind dovetail joint is an enduring symbol of fine craftsmanship. Every proud woodworker who has conquered the dovetail wants to show it off. In the old days, when every piece of wood was worked by hand, mastering this joint took lots of practice. But not today. Using machined parts and my method, you can make perfect half-blind joints the very first time you try. There's no fussy trial-and-error fitting.

Generally, a woodworker can follow one of two paths to make dovetails by hand. The classic artisan's method requires going for broke and sawing precisely on a line. It's fast and rewarding, but it takes a sure eye and a steady hand. The second, more cautious, approach allows you to saw away from a line, then pare to the line using a chisel. It's slower, but by guiding the chisel with a jig, anyone can do it. That's the method I'll show you here. The secret is to use very sharp chisels with specially ground sides (grind 12° bevels on both sides of your chisel and extend the bevels back ¾"), and stick to the directions.

My method. A few simple jigs guide your chisel; every paring cut is straight and square. It's not the fastest way to cut dovetails, but when you use it, I can promise you precise joints, even your first time.

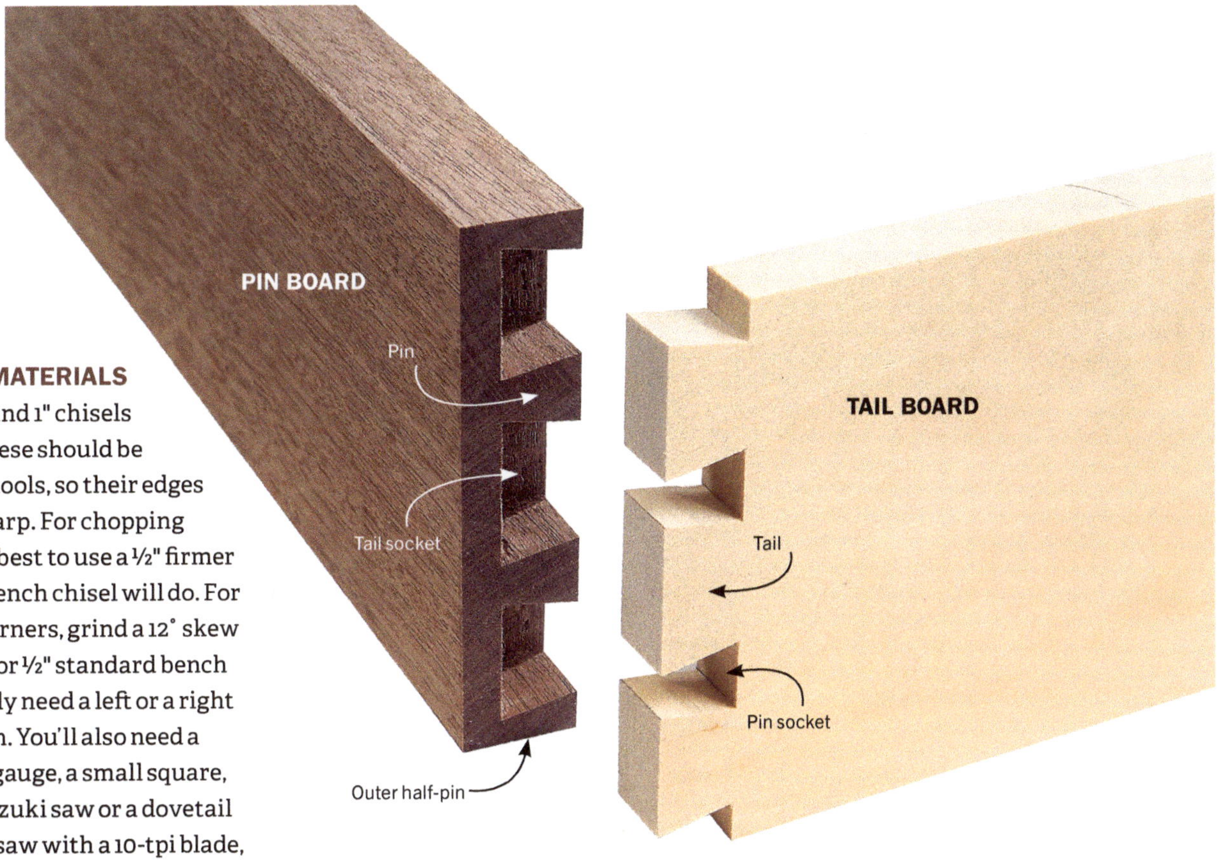

Precise Hand-Cut Dovetails | **HAND-CUT DOVETAILS**

TOOLS AND MATERIALS

You'll use 3/8" and 1" chisels for paring. These should be high-quality tools, so their edges stay razor-sharp. For chopping end grain, it's best to use a 1/2" firmer chisel, but a bench chisel will do. For paring into corners, grind a 12° skew angle on a 3/8" or 1/2" standard bench chisel. You only need a left or a right skew, not both. You'll also need a sliding bevel gauge, a small square, a Japanese dozuki saw or a dovetail saw, a coping saw with a 10-tpi blade, and a mallet or hammer.

Wood selection is important. For easy paring, the drawer front, or pin board, should be straight-grained and moderately dense. Walnut, cherry, and mahogany are excellent choices. The drawer side, or tail board, should be light in color and also easy to pare. Basswood is the best choice, but white pine and yellow poplar are good, too.

To start, machine each piece flat and square. Make the three guide jigs (p. 47). Orient the drawer side so the grain on its face runs toward the back. This makes the completed joint easier to plane flush. Sharpen your chisel, and let's begin. ∎

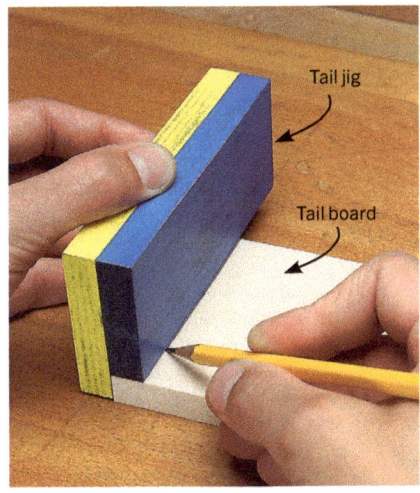

Draw a pencil line across the tail board. This indicates the dovetails lengths. Here, you'll use the first of three simple jigs. This is the tail jig. I've painted its components so they're easy to see.

Draw a pair of marks for each pin socket. Evenly space the sockets by eye. Make each mark 1/32" on either side of a 3/8" chisel. This ensures that the chisel will fit inside a completed socket. Draw marks for the outer half-pins.

HAND-CUT DOVETAILS | Precise Hand-Cut Dovetails

Transfer the dovetail angle. Take the angle from the jig to a sliding bevel gauge. The dovetail jig has two equal angles precisely cut on the table saw. This jig will guide your paring cuts in the final steps.

Lay out the tails. Draw long, fine lines through each pin mark to make triangles. Each one forms a tail. Refer to a sketch so the lines angle the right way. Flip the square as needed, so most of its handle always butts against the board's end.

Draw square lines across the board's end. This completes the tails. Shade the pin sockets between the tails to clearly indicate the waste areas you'll remove.

Clamp a thick, wide board to the tail board. Position it 1/16" above the pencil line you drew in the first step (p. 37, bottom left). Use a small square to check the distance. Raise the boards to a comfortable sawing height.

Precise Hand-Cut Dovetails | **HAND-CUT DOVETAILS**

Rough-cut the tails. Saw 1/32" inside each pencil line, within the darkened waste area. Keep the saw level and stop when you reach the guide board.

Remove the half-pin waste pieces at both ends of the tail board. Like the last saw cuts, these cuts are approximate. Saw level along the top of the guide board, being careful not to cut into the tails.

Remove the full-pin waste pieces with a coping saw. First, make one diagonal cut. Come back across, flush with the guide board, to release the waste piece. Now you're ready to pare the tails exactly on the layout lines.

Set up the tail board for paring. Place a piece of scrap plywood under it to protect the bench. Clamp the dovetail jig precisely along one of the angled layout lines. Support the jig, if needed, with another piece that is the same thickness as the tail board.

THE DOVETAIL BOOK | 39

HAND-CUT DOVETAILS | Precise Hand-Cut Dovetails

Pare the tail using a 1" chisel. It's easy to balance along the jig's side. The thinner the shaving, the better. To control each shaving's thickness, begin with one or more playing cards against the jig. Remove one card after each stroke.

Pare the last shaving with the chisel against the jig. For a super-smooth cut, work your way in from the end of the tail, taking one-third of a full shaving's width at a time. Reposition the dovetail jig on the other layout lines to pare both sides of each tail.

Set up the jig to pare each socket's end grain. First, loosely clamp the dovetail jig on top of the support board.

Clamp the tail jig to the dovetail jig. Slide the tail board to butt up against the tail jig. Place and tighten a clamp to the left of the tail board and remove the tail jig. Tighten the support board clamp.

Precise Hand-Cut Dovetails | **HAND-CUT DOVETAILS**

Pare the end grain. Use playing cards again to minimize each shaving's thickness. You may have to start with three or four cards. Drive the chisel with a mallet, if necessary.

Bear against the jig on the last cut. To pare into these acute corners, grind 12° bevels on both sides of your chisel. Extend the bevels back ¾". A 12° bevel is slightly steeper than the 10° dovetail angle.

Clean the corners by paring from the end. Then pare from above, as in the last step, to release the shaving. Your tails are now perfectly straight, smooth, and square to the board's face. They almost look machine-made!

Prepare the pin board. This step will make very fine layout lines easier to see. Coat the board end with a primer coat of shellac or varnish, followed by white correction fluid. Primer prevents the white fluid from penetrating into the end grain.

THE DOVETAIL BOOK | 41

HAND-CUT DOVETAILS | Precise Hand-Cut Dovetails

Clamp a third jig, the pin jig, to the pin board's outside face. Score a fine line across the board's end with the corner of the wide chisel. One light pass will do it. This is the baseline for the tail sockets.

Clamp the pin board level with the support board (p. 39, lower right). The pin board must be rock-solid for laying out the pins. Prevent the vise from racking by inserting a spacer that's the same thickness as the pin board.

Mark the pin board. Lightly push on a chisel butted to the side of each tail. This leaves a distinct, super-thin line. Butt the tail board's end to the scribed line. Use a square to align the pin and tail boards.

Clamp the tail jig to the pin board's inside face. Scribe another line with your chisel. This line indicates the pins' depth. After assembly, the pins will be flush with the tails, which makes gluing and clamping easier.

Precise Hand-Cut Dovetails | **HAND-CUT DOVETAILS**

Draw the pins on the board's inside face. Go well beyond the scribe line. The longer the lines, the easier they will be to follow when you saw. Shade or mark Xs in the waste areas, or tail sockets, between the pins.

Saw the pins. Stay 1/32" inside the lines. Stop short of both scribe lines. Clamp the pin board diagonally in the vise so you can see both faces. This enables a pull saw to cut smoothly, because it's cutting with the grain.

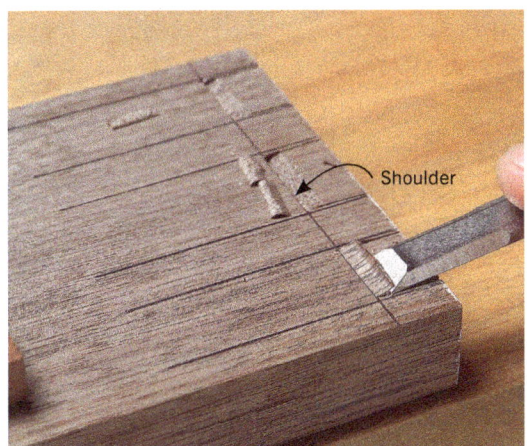

Pare at a shallow angle. This creates small shoulders in each tail socket. Clamp the board down, push almost to the scribe line, twist, and lift out a thin chip. These shoulders will guide the chisel in future paring cuts.

Chop the tail sockets. Hit the chisel once, straight down, 1/32" away from the shoulders. Switch to a 1/2" firmer chisel for this heavy-duty work. Reserve your bench chisels for paring. Sharpen the firmer chisel at a durable 35° angle.

HAND-CUT DOVETAILS | Precise Hand-Cut Dovetails

Make chips! After each downward chop, lower the chisel (bevel down) and split off a thick chip with a single blow into the end grain. Twist the chip to pry it loose. Continue chopping downward and sideways until you're within 1/16" of the lower scribe line.

Finish chopping the shoulders right on the scribe line. Rest the chisel against the small shoulder (p. 43, bottom left), hold it plumb, and strike one blow. By taking a thin shaving, you'll make a crisp, clean, deep shoulder.

Lean the chisel a few degrees on the second blow. Continue at that angle to the bottom of the tail socket. This traditional undercut ensures the joint goes together without any gaps along the scribed line.

Pare to the scribe line. Clamp the pin jig to the pin board once more, and use the playing card technique to remove thin shavings. Pare into the corners with the bevel-sided chisel.

Precise Hand-Cut Dovetails | **HAND-CUT DOVETAILS**

Remove the shavings. Cut them off at the base. Push hard and give a little twist; they'll pop right out. Don't bother for now with the shavings stuck in the corners.

Pare the pins with the dovetail jig. Clamp it to the pin board. Align one side of its notch with one of the saw cuts to remove a thin shaving with the wide chisel. This will be a sneak-up-to-the-line approach, without using playing cards.

Make the last paring cut. It should be right on the layout line. This method is so precise that you can split a hair, which is about the thickness each of these layout lines cut into the paint. Pare all the pins this way.

Tap the dovetail jig after each shaving. This shifts the notch closer to the pin layout line. Pare again. Each shaving should be very thin, so you can easily make an accurate cut. You don't have to loosen the clamp, because the jig shifts so little.

THE DOVETAIL BOOK | 45

HAND-CUT DOVETAILS | Precise Hand-Cut Dovetails

Pare the corners from above. The chisel's beveled edge lets you get right into this tight angle.

Clean out the corners with a homemade skew chisel. Push the chisel into the corner, twist it, and pop out the shavings. For a right corner, flip over the chisel and use it bevel down. This tool does the job very quickly.

Test the joint's fit. The pin and tail boards should go together without any additional work and end up flush. If seating the joint requires more than light hammer taps, use the dovetail jig to pare the pins that are tight.

Even up the joint after it's glued. It only takes a light sanding or a few plane strokes to remove the white paint and make the pins flush with the tails. Now you'll see how tight the fit really is!

THE 3 JIGS

TAIL JIG

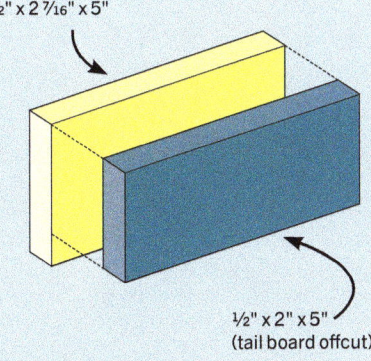

½" x 2 7⁄16" x 5"

½" x 2" x 5" (tail board offcut)

The tail jig. The secret to success with these jigs is to use actual pieces from your drawer. The blue piece below is an offcut from a drawer side, or tail board. In drawer making, a tail board is typically ½" thick, but any thickness will work with this system. Glue or nail the pieces together with brads.

PIN JIG

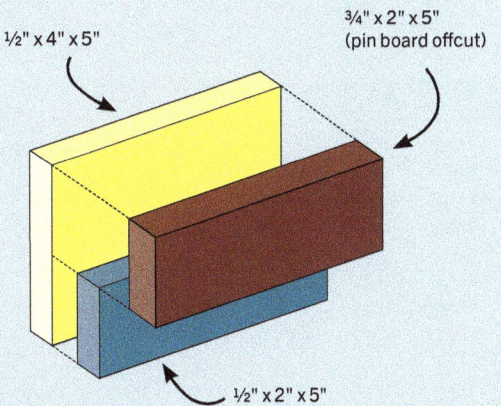

½" x 4" x 5"

¾" x 2" x 5" (pin board offcut)

½" x 2" x 5"

The pin jig. The accuracy of this jig is also based on using offcuts. The red piece is an offcut from a drawer front, or pin board. In drawer making, a pin board is typically ¾" thick, but any thickness will work. The blue piece is a second offcut from a tail board.

DOVETAIL JIG

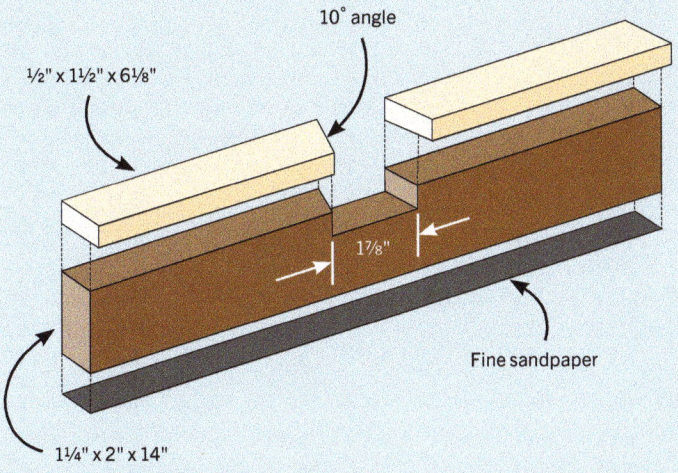

½" x 1½" x 6⅛"

10° angle

1⅞"

Fine sandpaper

1¼" x 2" x 14"

The dovetail jig. After gluing the parts together, cut the angles in this jig on the table saw. Check the cuts with a sliding bevel gauge to make sure the angles are exactly the same. Glue a piece of sandpaper to the jig's bottom. This prevents the jig from moving when you clamp it to a workpiece.

TIPS TO AVOID DOVETAIL GAPS

Here are a couple of tips for your dovetail toolbox.

BY CHRISTOPHER SCHWARZ

You've cut all your pieces and are putting everything together when you first notice it—a gap. A dark void where there should be none. Don't panic—it happens to the best of us. For whatever reason, there are instances when your joints just don't fit perfectly. These tips will give you a couple ideas on how to avoid unsightly, embarrassing gaps.

COMPRESSION MAKES DOVETAILS TIGHT

Hand-cut dovetails are some of the most challenging joints to fit perfectly. Many woodworkers will spend hundreds of dollars on router jigs or woodworking classes to get an airtight fit. If you decide to hand-cut your dovetails, there are a few ways to make sure you get it right.

Because wood is—on a cellular level—similar to a bunch of soda straws glued together, you can compress it a little bit. Usually, compression is a bad thing, such as when you drop a hammer on your work and it dents. But a little bit of compression is good when dovetailing.

Here's how it works: Cut the first half of your joint as you usually would—I usually cut the tails first. Then use that first half to knife in the second half of the joint—in this case, the pins. Next, when you saw your

Tight joints. With a little help, you can tighten ill-fitting joints.

pin lines, don't saw right up against the knife line you marked, as most books tell you. Instead, saw slightly wide. How wide? The whisker of a gnat would be a good place to start. Here's how I do it: After I knife in my joint lines, I run a pencil over each knife line. Then I start my saw cut to leave the entire pencil line.

Like all things pertaining to dovetails, this takes practice. Cut some sample joints to get a feel for it and use a magnifying glass to gauge your progress.

Once you cut your pins, use a knife to ease the inside edges of your tails, which will be inside the joint. When you join your two pieces, the too-tight pins will compress the tails and the joint will be seamless. If you try to compress too much, one of your boards will split as the two boards are knocked together.

Tips to Avoid Dovetail Gaps | HAND-CUT DOVETAILS

This compression works especially well with half-blind drawer joints where you are joining a secondary softwood for the sides (such as poplar) with a hardwood drawer front (such as oak), because the softwood compresses easily. But be careful: This trick doesn't work when you are trying to join two pieces of dense exotic wood, which doesn't compress much at all.

FAKE HALF-BLINDS FOR DOVETAIL JOINTS

Half-blind dovetails are trickier to cut than through-dovetails, but they don't have to be. I picked up this trick from dovetailing maestro Rob Cosman.

Essentially, you first build a drawer with the easier through-dovetails and then glue a ¼"-thick piece of veneer over the drawer front, making them look like half-blind dovetails.

Usually, with drawers, you have ½"-thick sides and a ¾"-thick front. To do what we're suggesting, make your drawer front with ½"-thick stock, too. Join the sides to the front using through-dovetails. Then, using your bandsaw, resaw a piece of ¼"-thick veneer from a piece of really nice figured wood. Make it a little larger than the finished size of your drawer front. Glue that veneer to the drawer front, let the glue dry, and trim it flush. This makes excellent half-blind dovetails and allows you to stretch your supply of nicely figured woods for your drawer fronts. It also works if you have a little bit of a gap in your joint that you want to hide. ■

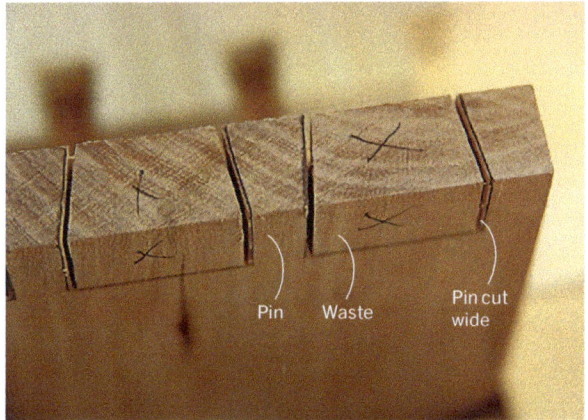

Wood compresses. You can use the natural compression of wood to make your dovetails tighter. Cutting your pins just slightly wide will force them to compress the tails.

Through-dovetails to simplify. To make life easier (and to stretch your stock of valuable wood) cut through-dovetails when joining your sides and drawer fronts instead of half-blind dovetails. Then add ¼"-thick veneer to the drawer front.

Attach the drawer front. Use your workbench as part of your clamping setup when applying the veneer to the drawer. This setup helps spread pressure evenly across this large surface.

THE DOVETAIL BOOK | 49

HOUNDSTOOTH DOVETAILS

Add strength and beauty to any corner with this dovetail variation.

BY ROB COSMAN

I first saw this joint illustrated in *The Encyclopedia of Furniture Making* by Ernest Joyce. I was fascinated by the complexity of it and for years wished I could cut them myself.

When I finally cut the joint, I realized the only difficult part was the initial layout; the rest was a matter of sawing and chiseling to the line. I have since been teaching others to cut them and most would agree that it is not as hard as it looks. If you have mastered the through-dovetail, this is like cutting two of them at once. There are some tools that will make the job easier. In addition to the proper tools, there are a few tips I have picked up that should help the novice—and maybe the professional.

WHY USE THE JOINT?

In considering strength, any well-done dovetail will usually be stronger than the application requires. However, the houndstooth adds more long grain–to–long grain glue surface, as well as increasing the amount of pin wood in the joint. In a through-dovetail, adding pins to create extra glue surface can make the baseline of the tail board the joint's weak spot.

However, with two scribe lines, one at the base of the small tails and one at the base of the large tails, the tail board strength is not compromised with the houndstooth. For this reason it could be considered the best way to make a strong corner even stronger.

THE TOOLS

Choosing the right tools for cutting dovetails will result in less frustration.

In short, you need a dovetail saw with a straight blade and little set to the teeth. I prefer a rip-tooth configuration. You also need a well-tensioned fret saw, a sharp marking gauge, dividers, a marking knife, a device for marking the slope of your tails, and chisels with small side bevels that allow you to get into the tight corners of the tail section of the joint.

MARKING THE TAIL BOARD

I cut all my dovetails' tails first. I begin by handplaning the surfaces of each piece flat, smooth, and square; this is the first step to an accurate joint and it makes it easier to see pen and knife marks. I label the face side of each piece (my guide here is simply to put the best face out).

I set the marking gauge for the exact thickness of the pin board and scribe this dimension all around the tail board.

After teaching hundreds of people to cut dovetails, I found the biggest problem (after sawing technique) was

Cut a shallow rabbet. This rabbet on the tail board will help you mark your pin board later. The pin board will drop into the rabbet, which will increase your speed and accuracy.

Half pins. Use dividers to lay out the location of the half pins on the tail board. Mark a half pin on one edge as shown; then repeat the process on the other edge.

Mark the tails. Use your dividers to step off three tails as shown (and marked in ink). Then reset your dividers to divide those tails in half for the houndstooth.

HAND-CUT DOVETAILS | Houndstooth Dovetails

the transfer of tails to pins. To make this easier and more accurate, I use a technique borrowed from Ian Kirby. I use my skew block plane to cut a shallow (1/32" to 1/64") rabbet on the inside edge of the tail board. This shallow rabbet provides perfect registering for the tail board when it is set on the end of the pin board prior to scribing the pins.

The skew block is convenient to use because of its built-in fence. I set the fence so that with it riding against the end of the tail board, the far point of the blade (set to clear the edge of the plane) cuts right on the scribe line. One or two passes will usually do; however, if the wood is prone to tearing in this cross-grain cut, then a few shallow passes are more effective than one heavy cut. Most important is to keep the rabbeted surface parallel to the inside face of the tail board. The next step is to set the marking gauge to establish the baseline of the small houndstooth pins. I find it best to start with houndsteeth that are two-thirds the thickness of the pin board. With this setting, I scribe the two faces of the tail board; then, with the same setting, I scribe the end of the pin board referencing off the face side. I now set the gauge to be 1/32" less than the thickness of the tail board at the rabbeted end and scribe this on the two faces of the pin board.

I use a pair of dividers to establish the width of the half pins in the end grain of the tail board. I want these half pins to be thick enough to prevent them from breaking or splaying during assembly. On a 5"- to 6"-wide joint, I would make the half pins about 1/4" at the top. I use one leg of my dividers to register on the outside edge of the tail board and I make a point in the end grain with the other leg.

With the interior pins just the width of my saw blade, I use another pair of dividers to step off the number of large tails I want between the half pins. Starting on one half-pin point, I walk to the other half-pin point using the number of tails as the spacing. I adjust the width of the dividers one

Crosshatch the waste. You will save yourself a great deal of frustration if you mark all your waste lines with crosshatching.

Begin to saw. Make sure your sawing motion is in a straight line from the saw tip to your shoulder.

way or the other until they land right on the opposite half-pin point.

Until you have the exact setting, be careful not to leave any points in the end grain. To lay out the small houndstooth tails, I split each of the large tails in half (dividers make this easy). Then, using a dovetail marker, I mark out the tails. With my pen in the first half-pin point, I slide the marker over to it and strike a line across the end and up the face from the second (or farthest) scribe line. On the second point, I strike a line across the end and up the face starting at the first (or closest) scribe line. The third point starts at the second scribe line, the fourth at the first, and so on until I reach the second-to-last point. On the last point, I turn the gauge around to mark the opposite slope of each tail, the first one starting at the second scribe line, the next one at the first, and so forth. I also take the time to clearly mark the waste. A moment invested here can save frustration and expense as a result of having to start over because of an errant cut.

CUTTING THE TAILS

The tails are then ready to cut. I secure the tail board in the vise so it's plumb and keep it low to reduce sawing vibration. I use a shoulder-wide stance and my right foot makes about a 60° angle with the front edge of the bench.

I grip my dovetail saw loosely with three fingers, index finger pointing down the saw. I want my sawing motion to be in a straight line from the tip of the saw to my shoulder. Think of the drive piston on an old steam engine for this. I use my left thumb and forefinger as an anchor point to start my saw. I pinch the top of the board with the bottom third of my thumb and forefinger. This keeps the end point of each digit above the set of the saw teeth. Light lateral pressure with the saw against my thumb and finger will ensure the saw starts cutting where I want it to rather than skirting across the end of the board.

With my thumb and finger pinching the end of the board, I can "inch-

Waste removal. After you define your tails with your dovetail saw, remove as much waste as possible with a fret saw.

Half-pin waste. Sawing off the half-pin waste on the ends of the tail board is a critical operation. Make sure your layout lines are precise and deep.

HAND-CUT DOVETAILS | Houndstooth Dovetails

Chisel the waste. Remove the waste between the tails. Notice the body position. This allows me to ensure my chisel is indeed vertical when chopping.

worm" them one way or the other to move the saw closer and parallel to the line. Lifting to take most of the weight of the saw gives me a smoother start. I start my cut on the forward stroke; once the kerf is started, the weight of the brass back will supply the needed downward pressure. I do add a little extra to speed things up. My job is to aim the saw and move it forward and back. One of my students came up with the five Ps to sawing: 1) Pinch the wood; 2) Press the saw against the fingers; 3) Position the blade against the line; 4) Pull up on the saw, taking most of the weight; 5) Proceed with the cut.

It can help the new sawyer to first get a shallow perpendicular kerf started across the end grain, then pause, aim the saw to match the angled line, and begin sawing. Don't try to correct an errant cut. It is better to continue off a few degrees and have a straight cut than to mess up the kerf by trying to change direction halfway in. Developing the skill to hold the saw level during the cut comes with practice; in the beginning, saw to the face scribe line, then carefully tilt and saw to the inside scribe line. Cutting to the line will help when cleaning out the waste between the tails—particularly between the small houndstooth tails. I make all the cuts angling the same way before changing and coming back the other way. This gives me a better chance of getting the angles right.

REMOVING WASTE

Before I reposition the board to cut off the half pins, I use the fret saw to remove the bulk of the waste between tails. I slip the blade down the kerf to the bottom; feeling the bottom of the kerf with the fret saw blade gives me an idea of level. I then rise up just a bit and begin sawing as I turn the blade. The closer I saw to the scribe line with the fret saw, the easier my chiseling will be. I take care not to cut into the side of the tail.

My next step is to orient the tail board horizontally in the vise to cut off the half pins. This is the first critical cut. The shoulders have to be on the mark. When I cut out the tails, I am creating a template to which pins will be made to fit. However, this cut already has a template—the inside face of the half pins. The inside of the half pin must meet exactly with the scribe line to ensure there's no gap.

The marking gauge line plays a big role here. If it has been scored deep and clean, as I start to move the dovetail saw across the wood on the waste side of the line, any material between the saw and the line will disintegrate. The saw will magically slide over to the shoulder line, providing a reference point for the saw. Now all I have to do is continue to saw vertically and be careful not to scar the side of the tail as I finish the cut.

CHISELING BETWEEN TAILS

With the tail board flat on the bench, it is time to chisel out the waste. I always start chopping from the inside of the board and finish from the face side. This is always a two-step process, half from each face. Because the final chop eventually breaks through the waste, should the chisel get away from me the damage will be confined to the inside of the joint. I always use a backer piece to protect my bench.

While some authors advise undercutting this part of the joint, I instead hold my chisel vertical to avoid two problems. The first problem with undercutting is the possibility of exposing a gap, should more face-grain material need to be removed than was planned for. These gaps can be fixed but are hard to disguise. The second problem comes from what I call "push back." The wedge shape of the chisel creates a fair bit of pressure from the waste side of the chop. If the chisel is angled as in undercutting, the wood left supporting the chisel edge at the scribe line is a mere point. Holding the chisel vertical provides the maximum amount of wood at this critical spot. Also, the more waste there is, the greater the pressure on the chisel. For this reason, I try to saw as much of the waste away as I can.

If I have sawn to the line on both sides of the tails, my corners will be clean. If not, I use the corner of a narrow chisel to clean them up.

Cleaning up the bottom between the small tails is a bit more difficult because of the limited space. With the waste removed, I use a wide chisel to clean the outside half-pin corner. This spot usually has a bit of material left because the square-bottomed saw can't cut into the angled corner.

STARTING THE PIN BOARD

The next step is "make or break"—transferring the tails to the pin board. I like to clamp the pin board upright in the shoulder vise, setting the business end flush with the top of my smoothing plane set on its side. With the pin board firmly held, I move the plane back 7" or 8" so I can create a bridge between the two using the tail board. I reference the edge of the tail board rabbet against the inside edge of the pin board. I make sure the long edges of the tail and pin boards are flush. The downward pressure I now apply with my left hand on the tail board is transferred to a much smaller surface area (the end of the pin board and edge of the plane), which makes it easier to keep the tail board from moving

Small bevel-edge chisel. A smaller chisel allows me to remove any additional waste between the tails.

Transfer layout. Position the tail board rabbet on your pin board. Use a marking knife to transfer the tail board layout onto your pin board.

HAND-CUT DOVETAILS | Houndstooth Dovetails

while I mark the pins. I approach the side of the tail with my marking knife on a 20° angle. As I bend the knife to lay the blade flat against the side of the tail, the force of the angle holds it tight while dragging the point through the end grain of the pin board. I explain this to my students this way: 80 percent of your effort and concentration needs to be on keeping the knife tight to the side of the tail; only 20 percent is on the mark being cut in the pin board.

Scribing the small tails takes even more care; the space is small and keeping the knife tight to the tail is difficult. Once all the marks are made and before I remove the tail board, I use my bench lamp to inspect the scribe lines to make sure none wandered away from the sides of the tails.

As a final procedure on the tail board, I chamfer the inside edges of each tail. This is done with a chisel and I start the chamfer in about 1/16" from the end. Although it may change depending on the thickness of the piece, I usually run the chamfer an 1/8"-wide down the side of the tail. I use the chisel to clip off the pieces where they meet the scribe line.

SAWING THE PINS

Before I start work on the pins, I make sure the board is standing plumb in the vise. The easiest way to check this is with a square against the benchtop and the board. Because I make my vertical cuts by feel, it is critical that the work is plumb. Developing this ability is probably the best sawing skill to have. It helps if you own a pistol-grip dovetail saw because it will register in your hand the same way each time you pick it up. Round-handle saws don't offer this advantage. I have my students practice making multiple vertical cuts 1/16" apart in the end of a plumb-standing board. Doing this helps spot any left or right drifts that can be corrected.

To lay out the vertical pin lines, I find it easier to draw starting from

Chamfer the inside edges of your tails. This step will make your joint assemble more easily. I use a chisel.

Pin board waste. Mark your waste areas on your pin board and saw out the waste—trying to split the knife line in half with your saw.

the scribe than trying to stop on it. For this reason, I strike my lines from the scribe up to the knife marks. I use my dovetail marker on its side to do these. I know from experience it is easy to be so focused on sawing to the line that if the line was drawn past the scribe, I could accidentally saw past it as well. Because of this, I make sure my lines stop where I want my saw to stop.

At this point, it is important to mark the waste clearly and carefully. I take the time to draw hatch marks on all the waste; this is my best bet against cutting on the wrong side of the line, because knife marks are not that easy to see in the end grain of some woods. While some authors advocate rubbing chalk dust into or dragging a pencil through the knife mark to highlight it, I find these suggestions make it harder to be precise. I use my bench lamp to shine light across the end of the board, creating a bit of a shadow in the knife marks, which makes them easier to see.

The knife mark leaves a V in the end of the pin board; half of the V must remain with the pin and the other half is wasted in the kerf. If any of the pin half of the V is removed, the joint will be loose. If any of the waste half of the V is left, the joint will be too tight. In drawer work where two different species of wood are used, there can be some leeway.

Usually the secondary wood (drawer sides, internal components) is a mildly hard species. The primary wood is often the harder and denser of the two. In this case, the mild wood will compress a little, allowing for a good fit even if some of the waste half of the V is left attached to the pin. In casework where both pieces are of the same species, that same excess would cause a split in one of the components. Knowing how much you can leave with different woods can only come with experience. For this reason I find it best to shoot for perfection with each cut. Now using the same five Ps of sawing, I split the knife mark.

Having the small pin scribe line across the end of the pin board can be a bit confusing, so as soon as I have made all my vertical cuts, I carefully saw out the waste with the fret saw. I then use the fret saw to undercut the waste portion at the back of the small pin. I use a chisel to vertically pare away most of the waste. This makes it easier to tell what is waste and what stays.

CHOPPING THE PIN BOARD

With the pin board facedown on the bench, I chop the waste and am careful to chop in the scribe and down almost to the back of the small pin. I

Extra waste. Once you have removed most of the waste with your fret saw, you'll need to remove an extra chunk of waste behind the houndstooth by chopping and paring it out.

flip the board over; then with a chisel narrow enough to get between the large and small pins, I chop from the scribe until I break through the waste. It helps to tilt the chisel to ride the angled edge of the pin; this produces a clean corner between the side of the pin and the bottom of the socket. It also means less work in the next step.

Then I clamp the piece vertically and remove any waste. I want the corners to be clean and sharp. I am also careful not to leave a bump in the bottoms of the sockets. The final procedure requires that I turn the board around to finish paring to the back of the small pin. A narrow chisel is easier to use here and requires less force to cut the end grain. I first make a vertical pare with the chisel in the scribe at the back of the small pin, with the bevel facing the waste. If the grain is angled in the wrong direction for a clean pare, I take shallow pares to sneak up on the scribe. A quick horizontal clipping of the pared wood at the base, and the joint is ready to assemble.

DON'T DRY FIT THE JOINT
If you've worked accurately, dry fitting will only make the final fit less than it could have been. As stated earlier, I start with stock that is square, flat and smooth. Accurate layout lines allow for accurate sawing. The tail cuts have to be perpendicular across the end and straight to the scribe; exact angles aren't critical.

Transferring the tails to the pins must be accurate, and inspecting before moving the tail board will let you know if the lines in the pin board are where they should be. The pin-saw cuts must be perpendicular to the end and must split the knife marks. It is easy to see if you have left too much of the knife mark. You can check the bottom of the pin sockets with a straightedge to be sure there are no bumps that would keep the joint from seating. Providing the chisel work has been done to the scribe lines, this joint will fit. Everything a dry fit would tell you can be read from your cuts and scribe lines before assembly.

TIPS FOR GLUE-UP
Having everything handy makes the glue-up less stressful and all but ensures a successful finale: steel hammer, small square, pallet knife, glue, wiping rag, pounding block (an extra one is wise), and a few clamps just in case. I apply the glue sparingly; you don't want a lot of excess glue running down the boards, nor do you want it collecting in the bottom of the pin sockets.

I glue all the mating long-grain surfaces and a touch on the end-grain shoulders of the tail board. The wet glue will act as a lubricant to help you to get the joint together. The extra surface area in a houndstooth adds substantially to the assembly friction, so any help is welcome. With the pin board firmly in the vise, I get the joint started with just hand pressure.

Once things are lined up, I use a pounding block and hammer to completely seat the joint. The pounding block needs to be wide enough to cover the entire joint; this will prevent splitting the tail board should part of the joint go together more easily than another. I always orient the block so that I'm pounding through the end grain. This offers a much more positive transfer of force. It is a good idea to have the end of the block smooth and square. You don't want any unnecessary dents caused by a rough block.

If the marking gauge was set for a little less than the thickness of the tail board, the joint will seat before the pounding block starts running into the ends of the pins. This is a lot easier than trying to pound around all those small pins to seat the joint. At this point I remove the piece from the vise and check it for square. If it has to be adjusted, I always reseat it when I am done. If it's necessary, a clamp across the joint—squeezing the half pins together—will help hold it square and seated.

I wipe away any excess glue from the outside right away; planing will do the final cleanup. Any glue on the inside is best removed once it has skinned over enough so it doesn't stick to everything in sight. If there are noticeable gaps, they are easier to repair while the glue is soft. These are always repaired in the end grain and from the side that will show the most.

Wedges cut along the grain and from the same species will almost disappear. Some woods are more forgiving with this procedure than others. Mahogany end grain is the best I have seen at absorbing a wedge. With a bit of practice, it won't take long before repairing gaps is something of memories. ■

Apply glue quickly with a pallet knife. You will also be able to apply glue in the right places. Coat all the long-grain surfaces.

Use a pounding block. It distributes your blows evenly across the joint.

THE ART OF MAKING DOVETAIL DRAWERS

A master cabinetmaker shares his approach to designing and fitting elegant drawers.

BY MARIO RODRIGUEZ

There are just a few things my partner at Philadelphia Furniture Workshops, Alan Turner, and I don't see eye-to-eye on. But how to make a good drawer isn't one of them. When building furniture, we probably spend as much time making the drawers as we do making the piece itself. But this is an aspect of furniture-making many woodworkers don't devote enough attention to. In this article, I'll explain and illustrate how I built a pair of drawers for a small writing desk.

Drawers bestow even the finest hand-crafted furniture with a utilitarian character by arranging, storing, and providing access to objects. Yet it's important that their number, size, and placement contribute to the purpose and appearance of a piece, not detract from it.

Drawers are expected to operate easily, without sticking or rattling around. Drawer bottoms shouldn't sag or be left rough-sawn, and joints shouldn't be sloppy. Drawers shouldn't be an afterthought, disappointing the viewer and diminishing the experience.

When open, a drawer should reveal craftsmanship and quality consistent with the rest of the piece. You wouldn't go to the trouble of reproducing a Philadelphia highboy, then fit its drawers with metal slides.

FEATURES OF A FIRST-CLASS DRAWER

No matter what type of drawers you build, there are several essentials to good drawer making:

■ *Good design.* Drawers should be consistent with the piece being built.

■ *Good material.* Use the best that can be obtained; whether it's solid or plywood.

■ *Careful measurements.* Your measurements must be exact; measure carefully, then double-check the figures.

■ *Good, sound techniques.* Basic skills on machines and with hand tools will produce crisp, clean work.

Details make the difference. The drawers ride on the rails of the web frame, and the guides restrain movement from side to side.

■ *Patience.* Take it easy; making and fitting a drawer will take time and may test your patience.

These drawers are supported by and ride upon a web frame consisting of two latitudinal rails (front and rear) and three longitudinal rails (two side, one center). Drawer guides were attached to the web to track the drawers into the openings at the front of the desk. When initially installed, each guide intrudes slightly into the drawer opening by 1⁄16". Later, the guides are carefully planed to allow the drawers easy and smooth travel.

The height of this desk is 29" below the top. I allowed 24 1⁄2" leg clearance, giving me 4 1⁄2" for top and bottom rails, and my drawers. The combined thickness of the rails is 1 1⁄2", so I had 3" for the drawers. That is a good height for most objects stored in a typical desk. These drawers will be almost 14" wide and 18" deep, which is also a good size. A drawer that is deeper than it is wide will operate easily without racking or sticking.

SELECTING MATERIAL

For the drawer fronts, select clean, well-behaved material—something

HAND-CUT DOVETAILS | The Art of Making Dovetail Drawers

mild and easy to plane. For this desk, I selected a single piece of mahogany, long enough for both drawer fronts and the center dividing strip between the drawers, then milled it to ¾" thickness.

For the drawer sides, carefully choose your stock. Quartersawn stock is ideal; it's stable, won't twist or warp, and is easy to plane. I carefully went through a stack of maple boards and selected the ones with the cleanest, straightest grain. Maple is a tight-grained, hard-wearing wood, ideal for drawer sides. Mahogany or oak are other good choices.

Many woodworkers make the mistake of using material that is too thick, which produces heavy, clunky drawers. This is a small desk and the drawers will hold small, lightweight objects and supplies, so milling the sides to thin dimensions maintained the delicate nature and scale of the piece. You should proportion your drawer stock to the piece. For instance, drawer stock for an 18th-century spice box or a contemporary jewelry box might be as thin as ⅛".

Hardwood isn't usually commercially available in less than 1" thickness, so I resaw my stock. This often means interior surfaces of the wood with different moisture content than the exterior will be exposed, which can cause some movement. If you resaw, anticipate some slight twisting or cupping, so mill extra stock and select the best for your drawers.

MILLING AND JOINING THE PARTS

After a preliminary milling, sticker the stock while you work on the rest of the project to help it acclimate to the shop environment. Then, as you approach your drawer-making, take the boards down to their finished thickness.

Rip the sides ¹⁄₃₂" narrower than the opening and leave them about 1" longer than necessary in case you need to recut the dovetails. And, if possible, orient the grain direction to make it easier to plane and fit the completed drawer later.

Drawer fronts should be ripped and cut to fit precisely into their openings, with barely a hairline gap all around. You should try for a tight, close fit at this stage. When fitting the completed drawer boxes, the drawer front can always be planed to achieve the desired fit and appearance.

Dovetails are regarded as the strongest and most attractive way to join two perpendicular pieces of wood with the grain running the same direction. I like to use half-blind dovetails to join the drawer sides to the fronts and through-dovetails for the side-to-back joints. They

Let it rest. If you resaw thicker boards for drawer material, sticker them and let them sit for several days to acclimate before milling the stock to final dimensions.

have a distinctive and attractive appearance. They also provide the added benefit of squaring the drawer during glue-up, often eliminating the need for clamps.

To lay out your dovetails, follow the 1:8 rule or just lay them out at 10°, with half tails at each end and two full tails centered on the remaining space. I first make a sheet-metal template that gives me a clear pattern (that can be reused) to mark onto the drawer sides.

After marking the tails on the drawer sides, scribe the dovetail baseline onto the ends of the drawer front about ¼" from the face. This amount makes for a good appearance and a strong joint. On the interior surface of the drawer front, scribe a line about 1/32" less than the thickness of the drawer side. That way, when the drawer box is assembled, the drawer side will sit proud of the drawer front, which allows the side to be planed flush with the end of the drawer face, without altering the size of the carefully fitted drawer front. This also makes glue-up easier because you won't have to worry about damaging the delicate pins.

When cutting your tails first, you can cut directly to the line, remove the waste at the baseline, and not fiddle with lots of tedious clean up. If the saw drifts a little, that's OK—as long as the kerf is narrow, clean, and straight. Any deviation from the scribed outline can be transferred over to the pin board. Remember: Half-blind dovetails are only seen from one side, so the parts can be undercut and relieved to ease the fit without compromising the appearance or strength of the joint.

Proud sides. Cutting the dovetails to leave the sides extended from the drawer front simplifies fitting and maintains the smallest possible gap between the drawer and carcase.

A fitting start. Fitting the width of the drawer side to the height of the opening before building the drawer gives more control of the process.

Wasting away. With thin drawer stock, a jeweler's saw or coping saw will remove the waste between tails quickly and without the risk of damaging the work by chopping between the tails with a chisel.

HAND-CUT DOVETAILS | The Art of Making Dovetail Drawers

In the groove. The slot for the drawer bottom is located in the space for the lowest tail in the drawer front.

Smooth transition. The angled top of the slip bridges the corner between the sides of the drawer and the bottom.

Slip sliding away. Glued in place, the slip strengthens the lower portion of the drawer bottom. The back of the drawer is narrower than the side so that the drawer bottom can be slid into place after finishing.

Next, carefully mark the outline of the tails onto the pin board (drawer front). Cut the pins fat and pare them just shy of the line. The wide spaces between the pins will provide easy access from two angles. A partial test fit will reveal any excess material that has to be removed. I always say that folks won't notice a small discrepancy in spacing or angle in your dovetails—they'll only notice the gaps!

After fitting the drawer fronts to the sides and checking the dovetails for fit and appearance, cut a groove between the pins along the inside of the drawer fronts to accept the drawer bottoms. The placement of this groove is important for two reasons. First, it should be hidden when the drawer is assembled. Second, it should be situated to maximize your storage space.

THIN SIDES CALL FOR SLIPS

If you mill your sides to a thin dimension, you should consider using drawer slips for reinforcement and extra strength. Thin drawer sides might not be thick enough to support the drawer bottom, which is commonly held in place with a groove cut into the sides.

Drawer slips are small moldings, placed along the interior of the drawer sides to prevent their splitting under a load or heavy use. They also add character and detail.

I cut the slips on the table saw with a careful sequence of cuts that yielded the slender parts. I milled them to a profile that was adequate to support the drawer bottom, yet would only minimally intrude upon the usable space of the drawer.

The sides and back of the drawer are joined with through-dovetails. These can be a little tricky because thin stock will dictate small dovetails. And if the stock is very thin, it can split when you cut the joint. However, one small advantage is they won't be seen unless the drawer is fully withdrawn.

Before laying out the joint, measure the location of the drawer-bottom groove (from the drawer front) and cut enough material off the drawer back to allow the drawer bottom to be slipped underneath it when the drawers are assembled. Don't forget to cut off the excess length of the drawer sides. My sides-to-front dovetails turned out good enough that I didn't need a second chance at them.

After cutting the through-dovetails for the side-to-back joints, dry-fitting the drawer boxes, checking dimensions, and sanding the parts, I glued them up. If the dovetails are well executed, the drawers should come together without clamps. However, if you do need clamps, check the drawers for square. You should also check for flat by placing them on a flat surface. A twisted drawer will severely complicate the installation later.

When gluing up, keep a damp rag nearby. It's a lot easier to remove excess glue from the inside of the drawer at glue-up than to allow it to dry and have to chisel it out later.

When the glue in the drawer boxes is dry, measure and cut the drawer slips to fit against the inside of the drawer front and back. Then align the slips with the groove on the drawer front and the bottom of the drawer back; glue and clamp them in place.

When appropriate, use solid wood for the bottoms. For this desk, I cut a couple of 5⁄16"-thickness leaves from a 1"-thick piece and bookmatched them. After gluing up the panels, sand them to fair the seam and smooth the surface.

Taking the bottoms down to 5⁄16" kept them thick enough while making them lightweight, too. After cutting them to size, I rabbeted three edges. By rabbeting the edges I could slip them into a 3⁄16" groove, yet maintain their 5⁄16" thickness.

Because the drawers were narrow, I oriented the grain front-to-back, because I wasn't concerned with any significant wood movement. This

Match the groove. The rabbet in the drawer bottom matches the width of the grooves in the drawer sides and front. The extra thickness in the bottom keeps it from sagging.

Short and sweet. The drawer guides have a rabbet on the bottom and stop short of the back. This allows them to be planed during the fitting process.

means that grain shrinkage or swelling will take place across the drawer. When making your drawer bottoms and orienting the grain, take the size of the drawer into account. Generally, you want to run the grain in the direction of the longest dimension. So, a drawer that measures 16" deep and 24" wide would have the grain running side-to-side.

FITS LIKE A GLOVE

On a cabinet, I would remove the back before fitting the drawers. This provides easier access and the opportunity to "eyeball" any problems not visible from the outside. In this case, I left the top off the desk.

HAND-CUT DOVETAILS | The Art of Making Dovetail Drawers

When the drawers are ready, plane the sides flush with the front; be careful not to reduce the size of the drawer front. I plane a little more off the back end of a drawer, making the back slightly narrower than the front. This allows the drawer to initially enter the cavity with ease and tighten up slightly as it hits home.

I attempted to slide the drawer into its recess. But it was still too large to clear the opening. A few careful passes with a plane over the drawer sides fixed that. However, the box still wouldn't slide fully into the desk. With a sharp block plane, I took light shavings from the drawer guides (which were glued to the drawer web and flanked the opening) on each side. The guides are rabbeted and stop 2" short of the back apron, so their full length can be easily trimmed with a block plane. With several light, careful strokes, the drawers were planed to a tight fit.

Now the opening can be adjusted more precisely. By carefully planing the sides of the drawers, I was able to achieve a tight 1/32" gap all around the drawer front. But with the drawer resting on the frame, there was no reveal/gap along the bottom. This is one of the last steps in fitting a drawer and should be performed in a slow and careful manner.

To create an even gap, I scribe a line with a marking gauge and, using a tiny rabbet plane, I cut a small rabbet along the bottom edge of the drawer front. On a larger drawer, I'd use my shoulder plane.

Once the drawers are fitted, you can apply a small amount of wax to the bottom edges of the drawer sides and the guides. When you're satisfied with the operation of the drawers and their alignment with and to the front of the desk, you can set the drawer stops in place. These are three small blocks that are glued and clamped onto the front rail of the drawer web, just behind the drawer fronts.

Some woodworkers choose to leave the drawer interiors unfinished; others finish them exactly as the rest of the piece. It's a good idea to provide some form of protection. I recommend a light coat of shellac, lacquer or just wax. I finished these drawers with a very light coat of sprayed-on satin lacquer, rubbed out and waxed. ■

Rabbeted reveal. A small rabbet on the underside of the drawer front provides an even reveal to match the gaps at the top and sides.

Stop right there. Assembling and fitting the drawer before installing the drawer bottom allows precise placement of stops glued to the lower rail.

TAPERED SLIDING DOVETAILS

Hand tools are the way to go for this traditional joint.

BY FRANK STRAZZA

Tapered sliding dovetails are multipurpose joints traditionally used for drawer dividers, holding legs in place on a pedestal table, and attaching tops to case pieces.

The primary reason for a tapered joint (instead of a straight joint) is the reduced friction over its length. The tapered sliding dovetail gets tight only during the final fit, when it is ready to seat home. This reduced friction makes fitting much easier.

Once the joint is seated, it is extremely tight and often requires no glue. Hand-cutting this joint is easier and quicker than the lengthy process of setting up a jig and machine.

After incorporating tapered joints into much of my furniture, I've found them strong and multifunctional; I believe you'll enjoy their benefits, as well as the challenge of cutting the joint, as much as I have.

Below, I'll show you how to create a small tapered sliding dovetail, such as would be used for a drawer divider.

TAIL LAYOUT

For this exercise, I use two ¾"-thick x 4"-wide cherry boards (a typical size for a drawer divider).

HAND-CUT DOVETAILS | Tapered Sliding Dovetails

Tail layout. After marking the baseline at ¼", mark the taper on the end grain.

Tail cuts. I use a dovetail saw to make the tapered cuts; notice how I use thumb pressure against the saw to keep it correctly aligned to the cut.

Chiseled wall. Use a chisel to create a "wall" against which you can register your saw.

The first step is to cut the tapered tail (after making sure that the end of your board is perfectly square).

Set a cutting gauge (not a pinned marking gauge, which would tear the fibers rather than cutting them) to ¼" and scribe all around the end of the tail board to mark the baseline. On the board's edge, the sliding dovetail appears identical to a regular dovetail. However, from above, there will be a taper.

Place the tail board upright in your vise in preparation for marking tapered lines on the end grain. Starting with the end facing you, place a straightedge on the outside edge. Angle it toward the other end of the board, with a slope of about one-quarter the thickness of the board; in this case, that will be ³⁄₁₆". Draw a tapered line with a pencil along the straightedge. Repeat this process for the opposing side.

Using a dovetail marker or bevel gauge, draw the lines for the tail on the edge of the board. (I like a 1:7 angle.)

CUT THE TAIL

Cutting the angles on the top is a bit of a challenge, but no more than sawing a tenon cheek. Starting at the back, cut partway down your line, shift the saw to the front to make another partial cut, then join the two cuts. It is helpful to apply pressure between your thumb and the side of the saw to keep it aligned.

It's important that the two tapers on the end grain are straight and that the angles that make up the tail are cut accurately. Remember—you are only cutting down ¼".

Tapered Sliding Dovetails | **HAND-CUT DOVETAILS**

CUT THE SHOULDER

With a knife, deepen the shoulder line that you marked with your cutting gauge.

Now use a chisel to create a wall right against the knifed line (see p. 68, bottom). That creates a nice shoulder to set your saw against as a guide for accurate cutting. (This method is effective for extreme accuracy in cross-grain cuts.) But don't overcut! Inspect the tail to ensure that the shoulders are crisp and clean.

On occasion, I use a sliding dovetail plane to cut the tail, but I've found it not as effective for a narrow tail such as this one. (However, if you're creating a long sliding dovetail, such as on a tabletop, the plane is essential.)

MARK AND CUT THE PIN

The next step is to transfer the tail to the pin. Mark two pencil lines square across the pin board to indicate the width (¾") and placement of the tail board. Draw a face mark on the inside of your pin board on the right-hand side and another one on the right-hand side of the tail board. These marks should face each other; they will help you keep the alignment correct.

Lay your tail board on edge with the small dovetail end touching the pin board, with the end grain facing you. With your knife closest to the shoulder, mark the small end of the tail on both sides. Now flip the tail board end-over-end so the end grain faces away from you. Mark the large end of the tail on the edge of the pin board closest to you, marking both sides right at the shoulder.

Narrow, then wide. First, mark the small end of the tail (left). (Note that your knife position is critical here, and be sure to mark both sides.) Then mark the wide end on the other edge of the pin board (above).

Cordless router (below). Level the floor of the socket using a small router plane.

Waste removal. Work in from both sides to remove the waste. On the wider side, a ⅜" chisel is the best choice; working from the narrower side, switch to a ¼"-wide tool.

THE DOVETAIL BOOK | 69

HAND-CUT DOVETAILS | Tapered Sliding Dovetails

The key is transferring an image of the narrow section of the tail onto the pin board.

With a straightedge, join the marks for both tapers front to back. Then with a knife, lightly scribe along the straightedge. Remove the straightedge and go over the knife cuts several times to deepen the lines.

Transfer the 1:7 angles of the tail onto the edge of the pin board.

Using the same ¼" setting on your cutting gauge, mark the depth of the pin recess.

As you did for the shoulder cuts, use a chisel in your layout lines to create a wall for your saw to follow, then saw down the length of the taper line, carefully following the angle of the tail. Be sure not to overcut.

Working from both sides, remove the waste using a ⅜" and a ¼" chisel, with the bevels facing up.

With a small router plane, remove any excess material, bringing an even depth to the floor of the pin.

Now the moment of truth! Slide the tail into the socket and use a hammer to seat it tightly. If it's too tight and doesn't go all the way home, that's a good problem. Look at both ends, and you can usually tell where the problem lies. If it's too tight on one end, simply chisel away the material on the pin board. (If it's too loose, start over with a fresh pin board.)

And don't worry if your joint isn't perfect on the first try; it took me several practice sessions the first time, too. ■

Test. Now slide the tail into its socket; use a hammer to knock it home.

Truth. The finished sliding dovetail joint should seat firmly in place. If you've cut it perfectly, you don't even need glue.

Great divide. The tapered sliding dovetail is my joint of choice to divide casework drawers.

COMPOUND-ANGLE DOVETAILS

The system is as elegant as the joint itself.

BY TOM CASPAR

Compound-angle dovetails are some of the most beguiling joints in all woodworking. But, as John Lennon once suggested, they can be as difficult to make as "fixing a hole in the ocean." Well, maybe not that hard—but it all begins with figuring out those odd angles.

I've developed a system that makes layout easy. It's based on very simple and familiar geometry (see "The Basics", p. 72). It's universal, too:

HAND-CUT DOVETAILS | Compound-Angle Dovetails

THE BASICS

When it comes right down to it, there are only two things you have to know about compound-angle dovetails.

First, the ends of the pins (upper left) slant just like standard through-dovetails (upper right). It's the tapered side of the board that makes the angles look strange.

Second, the sides of the pins (lower left) are parallel to the top and bottom edges of the board, just like standard dovetails (lower right).

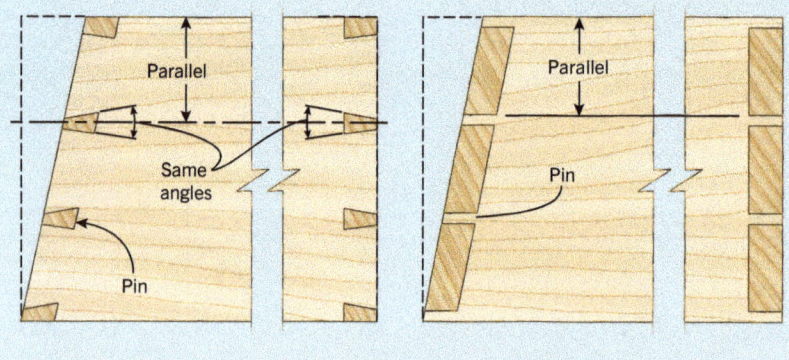

This system works for a project with any amount of splay and for dovetails of any pitch.

Before laying out the dovetails, your boards must be cut at the correct compound angle to form tight-fitting butt joints.

We'll lay out the pins first, rather than the tails. As you follow the photos, there are a couple of small details that I'd like to emphasize. First, be sure to draw the dovetail triangles (p. 73, top left) on the inside face of the pin boards. (If you draw them on the outside face, the angles you transfer to the edge of the boards will come out backward.) Second, always keep track of which side of your sliding bevel is facing: up or out. I use a piece of tape to mark one side, just to be sure. Now, let's start. ■

Draw a horizontal line. Place it across the inside face of one of the pin boards. The exact position of the line isn't important.

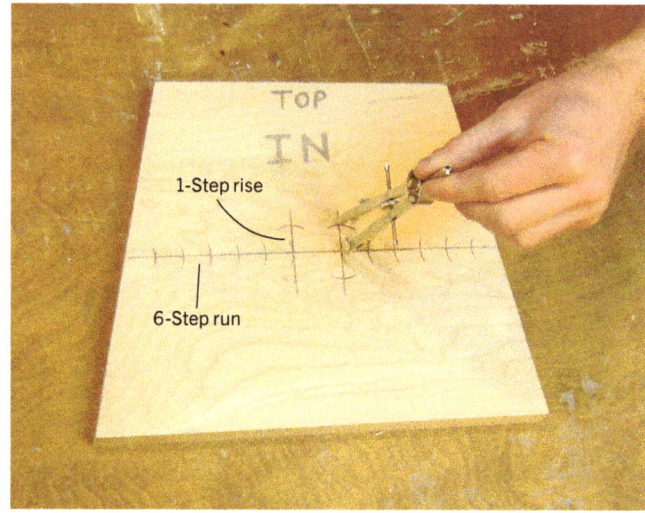

Rise and run. Mark the rise and run of the dovetails on the line. These dovetails have a pitch of 1:6, but you can use any pitch you wish.

Compound-Angle Dovetails | **HAND-CUT DOVETAILS**

Draw two triangles. Think of these two triangles as representing the ends of really large pins.

Get a bevel. Adjust a sliding bevel to match the right side of a triangle. Put a piece of tape on your square—this side should always face out or up in the next steps.

Mark the right sides of all the pins. If necessary, you can rotate the bevel to draw a line, but always keep its taped side facing up.

Reset the sliding bevel. It should now match the left side of the triangle.

Draw the left sides of all of the pins. Note that the line on the far right is actually the left side of a half pin.

Reset the bevel again. Now match the splay angle of the board.

THE DOVETAIL BOOK | 73

■ **HAND-CUT DOVETAILS** | Compound-Angle Dovetails

Draw the sides of the pins. Repeat the same process, starting with p. 73, top right, on the other end of the board.

Marking gauge thickness. Adjust a marking gauge to the width of one edge. Note that this is slightly longer than the thickness of the board.

Scribe. Use the marking gauge to scribe the shoulders of the pins on both sides of the board.

Comparison. Shade in the pins and compare opposing sides. If all the pins slant the correct way, start sawing!

IMPOSSIBLE DOVETAILS

Baffle your friends with these perplexing joints.

BY JOCK HOLMEN

Press a dovetailed board into another board with matching sockets, and you've created woodworking's most iconic joint. The dovetails and sockets wedge the boards together, so the joint can't pull apart; the only way to disassemble it is to lift the dovetailed board back out of the sockets.

So, what if you can't lift out the dovetailed board? How do you disassemble the joint? And how would you assemble this joint in the first place? Those are the questions to ask when you show a friend the dovetail joints shown here. These puzzling joints appear to wedge together in more than one plane—an impossible feat for traditional dovetails!

TEST YOUR HAND TOOL SKILLS

The secret behind these joints, of course, is that they don't assemble the traditional way. The first two are elaborate sliding dovetails and the last is a complex pivoting joint. There's no simple method to machine these joints; they must be cut primarily by hand. And creating them will test your hand-dovetailing skills, because of their compound angles and large joint surfaces.

Unlike most wooden puzzles, these joints shouldn't be constantly

HAND-CUT DOVETAILS | Impossible Dovetails

assembled and disassembled. The pieces include fragile short grain that can easily break and delicate edges that will quickly show wear. It's best to glue the joints together as soon as they've been satisfactorily fitted.

You're bound to make some mistakes, so always start by making a practice joint. Make sure to use stock that is straight-grained on all four sides—it's difficult to pare against the grain's slope. Also, it's a good idea to use hardwood for one piece and softwood for the other. This method is more forgiving, because the softwood piece will conform to the hardwood piece when you assemble the joint. Using hardwood for both pieces requires absolute precision, because there's no forgiveness: If the pieces don't fit perfectly, the short grain parts will simply break off.

Use the same steps you would follow to cut dovetails by hand to create all three joints. Start with pieces that are cut perfectly square. Lay out the dovetails and sockets on each piece. Clearly mark the waste areas. Make sure your tools are razor sharp. Cut the cheeks first. The safest method is to cut outside the lines. Next, remove the waste. Finish by paring to the line. It's best to scribe or knife the layout lines, so you can precisely bed your chisel for paring; if you pencil the lines, make sure they're crisp and narrow.

THE SECRET TO EASY ASSEMBLY

When all of the cheeks and shoulders of these joints are pared absolutely flat, it's difficult to slide the pieces together, due to friction resulting from the joints' large surface area. Fortunately, there's a work-around. The only places where the joints have to fit perfectly are the faces that show. So, to make the pieces slide together more easily, slightly hollow the joint surfaces that don't show.

DEVIOUS END JOINT

Through-dovetails appear to cross inside this joint, which, of course, is impossible. Instead, two dovetails run diagonally across the top of Piece A and two diagonal sockets are cut into the bottom of Piece B (see illustrations below). On each piece, the layout is identical on all four sides.

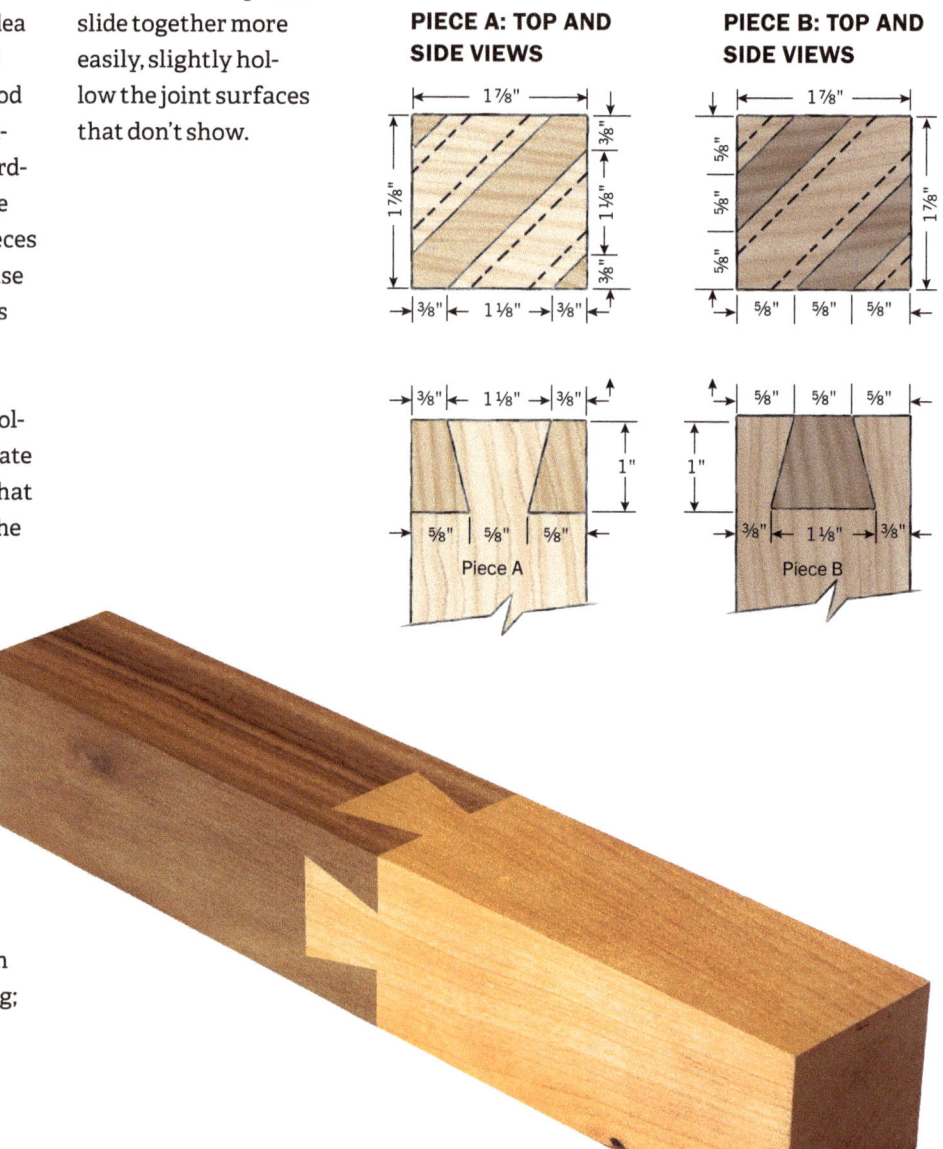

Impossible Dovetails | **HAND-CUT DOVETAILS**

Make the Pieces

Lay out the dovetails and sockets. Mark the waste. Saw the dovetail cheeks in Piece A and the socket cheeks in Piece B. Sawing these compound angles accurately is tricky, so don't be a hero: Cut wide of the lines, in the waste area.

Remove the waste to establish the joint shoulders. Insert the coping saw into one of the cheek cuts, turn the blade, and saw to the other cheek cut.

Pare to the lines. Use a wide chisel to pare the cheeks and a narrow chisel to pare the shoulders. Beveling the sides of the chisels makes it easier to get into the acutely angled corners.

Ease the Fit

On this joint, all four faces of both pieces show. To ease the fit, hollow each dovetail cheek on Piece A and the shoulder of each socket on Piece B. Always start paring 1/16" inside the outside edge, to create a lip. Then pare to the center. When the pieces slide together, the 1/16" lips at the outside faces will be the only parts of the joint that fit flush.

Cut the dovetail cheeks. The best strategy is to stay outside the layout lines.

Remove the waste. Use a coping saw.

Pare to the lines. Remove the excess material in several thin shavings. This requires a razor-sharp chisel and light, controlled pressure.

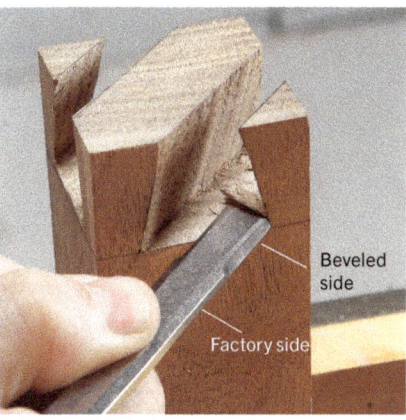

Cleanup. To pare cleanly into the angled corners, bevel the sides of your chisel.

Assemble. Slide the pieces together from corner to corner.

HAND-CUT DOVETAILS | Impossible Dovetails

DEVILISH LAP JOINT

On a typical lap joint, Piece A would simply press into Piece B. Well, that can't happen here. Neither can the two pieces pull apart. So what gives? A clever version of a tapered sliding dovetail, that's what. The dovetailed tenon on Piece A tapers on the bottom, from the shoulder to the end. On the top, its edges slope in the opposite direction, at compound angles. The mortise in Piece B mirrors the tenon on Piece B, sloping up on the bottom, and down and out on the top.

Make Piece A

Lay out the dovetail and mark the waste. Cut the 1/8" bottom shoulder on the table saw. Raise and tilt the blade to cut the tenon's angled bottom face.

Use a handsaw to crosscut the dovetail's square shoulders. Cut the dovetail's compound-beveled cheeks. Precisely pare the cheeks and shoulders to the layout lines.

PIECE A: FAR END, TOP VIEW, EDGE VIEW, AND CROSS-SECTION AT SHOULDER

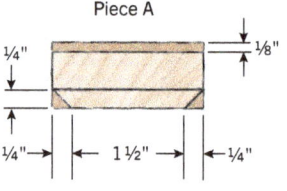
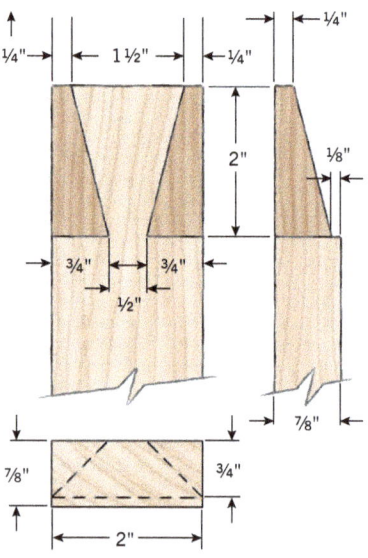

Make Piece B

Lay out the mortise and mark the waste. Saw the angled mortise shoulders, staying wide of the layout lines.

Make a lengthwise cut in the center of the mortise to divide the waste in half. Cut deep at the butt end and shallow at the open end, following the slope of the mortise.

Saw out the waste. Insert the coping saw in the lengthwise cut, turn the blade, and saw to one corner. Remove the waste and then saw to the other corner.

Pare to the lines. When you pare in from the butt end, the acute angles inside the mortise will trap the waste, so be prepared to progress slowly.

EASE THE FIT

Fortunately, only the top face and outside end of this joint show; the other hidden joint surfaces can be "adjusted." The sloped bottom face of Piece A and its beveled dovetail cheeks are the easiest surfaces to access. When you hollow these surfaces, however, do not disturb the narrow wedge-shaped end of the tenon, or the edges of its dovetail-shaped top surface.

PIECE B: FAR END, TOP VIEW, EDGE VIEW, AND CROSS-SECTION AT SHOULDER

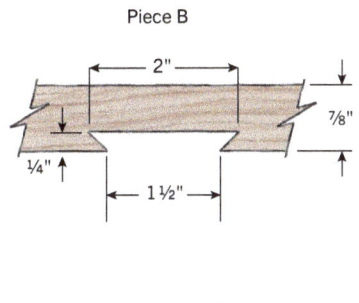

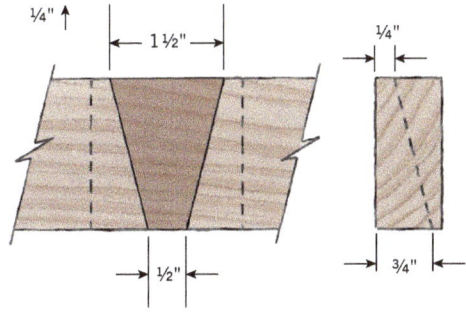

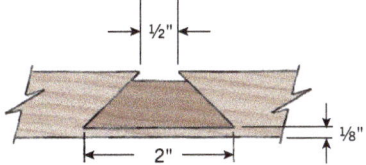

Assemble. Insert Piece A into Piece B. The dovetailed tenon is thin at the end, so it fits at the bottom of the mortise. As the tenon slides into the mortise, the dovetail on the face of Piece A rises until it's flush with the face of Piece B.

HAND-CUT DOVETAILS | Impossible Dovetails

DIABOLICAL CORNER JOINT

The flared ends of the dovetail pins mean this corner joint can't disassemble the traditional way. And no evidence of a sliding joint appears on the back side of the joint, so it can't go together like the double-dovetailed tenon in the previous joint.

The secrets are dovetails that slope at three different angles and sockets with coved shoulders (see illustrations below). They allow the boards to slide together in line and then rotate 90° to form the corner.

PIECE A: FRONT, EDGE, AND BACK VIEWS

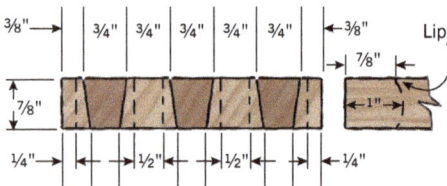

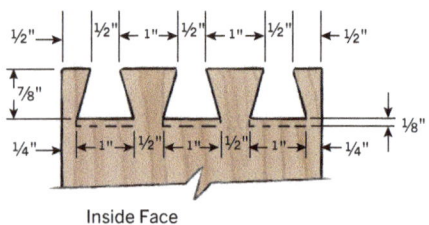

PIECE B: FRONT, EDGE, AND BACK VIEWS

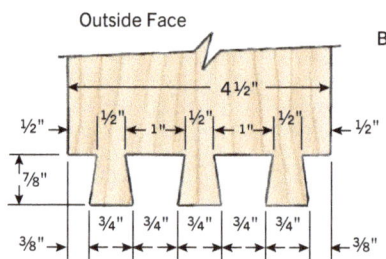

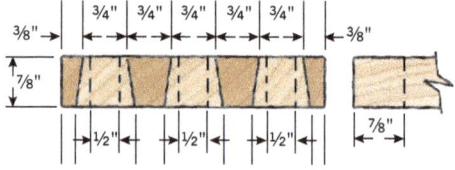

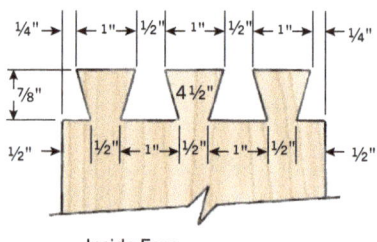

80 | THE DOVETAIL BOOK

Impossible Dovetails | HAND-CUT DOVETAILS

For the record, the slope of the dovetails on the outside face of Piece A matches the sockets on the end of Piece B, the slope of the dovetails on the end of A matches the sockets on the outside face of B, and the slope of the dovetails on the inside face of A matches the sockets on the inside face of B. The coved sockets in Piece A provide clearance for the outside corner of Piece B as the boards pivot.

Make Piece B

Lay out the dovetails and mark the waste. Saw the dovetail cheeks, following the layout lines on the end and the outside face. This cut won't come close to the dovetails on the back face because they slope more steeply. That's okay. Use the coping saw to remove the waste.

Pare to the lines. On the cheeks, work from each face to the center—on the back face, you have to remove more material. Because the front and back slopes differ, the cheeks' faces will be faceted, rather than flat (see illustration below). Notice that the dovetails on both faces are the same width at the bottom scribe line.

Make Piece A

Follow the same procedure used to cut Piece B, with this exception: Hollow out the socket shoulders, leaving tiny (1/32" wide) flat lips at the outside face to seat the joint (consult illustrations on p. 80 and to the right).

Ease the Fit

The cuts on the ends and outside faces of both pieces are the ones that show, so they must remain precise. To ease the fit, slightly widen the socket cheeks and shoulders on the back face of A—but do not disturb any dimensions at the end of the board. Gently ease the facet lines. Make sure each socket shoulder in A is hollowed into a fair curve, so the corner of B can rotate through. ∎

Assemble this joint in two steps. First, with both pieces oriented outside-face out, slide Piece B into Piece A from the back. When the pieces are flush, the dovetails on the outside faces don't fit.

Complete the joint. Carefully rotate the pieces to complete the joint. Bear the inside corners of A against the shoulders of B as you rotate.

3D VIEW

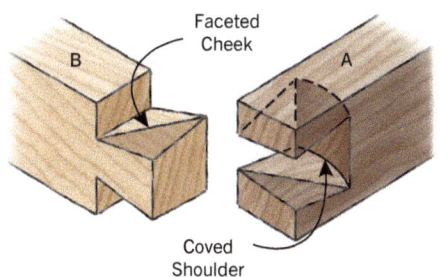

Faceted Cheek

Coved Shoulder

MITERED DOVETAIL BOX

A few key elements push your hand tool skills.

BY CHRISTOPHER WALKER

When my wife asked that I build her a box to contain a growing collection of aroma oil containers and associated accouterments for her piqued interest in aromatherapy, I was thrilled. Besides the house smelling great, I had spousal approval to build something—a win-win proposal.

Note that this project uses metric measurements. To convert from mm to inches, divide by 25.4.

DESIGN

The requirements were a single box with enough space to store current aroma oil bottles and any new items acquired. My wife purchased a drawer pull that caught her eye and asked that I incorporate it into the build. The final design was a single drawer carcase with a handle on top.

This project was a perfect opportunity for dovetail practice. The outer carcase design was well-suited for a first attempt at a mitered dovetail while the inside drawer was perfect to hone my half-blind dovetails. For the carcase, a groove was to be cut in to support a back panel. The bottom panel for the drawer was to be a panel set into a rebate.

LUMBER SELECTION

Black walnut is a rarity in Japan, where I live. It is imported from the United States and I was lucky enough to get my hands on some through an auction. This felt like the perfect project to use it. Prior to this project, I purchased several Japanese ash slabs with the intent of milling them up for drawer parts. The ash would provide a nice contrast to the dark tones of the black walnut while being softer, which would perform favorably when knocking the dovetails together.

CUTTING A MITERED DOVETAIL JOINT

Stepping up to the bench, a mitered dovetail can seem intimidating. However, if you have experience with a standard through-dovetail, you will be fine. With proper planning and a well-sorted layout, it becomes a simple matter of finessing away the waste.

Outside of the standard tool requirements for through-dovetails,

Draw it out. A design session led to the choice circled in green.

Stickered and stacked. Black walnut is a rarity in Japan.

HAND-CUT DOVETAILS | Mitered Dovetail Box

Visual cue. Blue tape gives a clear indication of where to cut, especially in dark woods like this black walnut.

Multiple gauges. Two marking gauges come in handy for marking the baseline and the miter.

you may also need a second marking gauge that will be set to the width of the mitered portion of the dovetails. A 45° chisel guide block cut to precision on a table saw or similar machine will also be useful.

To begin marking out the joint, set a marking gauge to the thickness of the material and mark the baseline all the way around the tail board except for the edge to be mitered. Then mark both faces of the pin board, skipping the edges, just like in through-dovetails. It is important to note that for this joinery method, the material for both tail and pin boards must be the same thickness.

Different thickness will create a misaligned miter.

You can reuse the same gauge setting to set the width of your mitered portion, but if the width of your mitered area will differ, use a second marking gauge. I like to keep all marking gauges set for a single purpose until the entire operation is complete. My mitered edge will be a panel that will be inset 15 mm from the edge. My material thickness is around 17 mm, which means two marking gauges are in order. To mark out the tail board, using the edge where the miter will be, score a line across the end grain followed by a line on the inside and outside face to the baseline. Then, with a bevel gauge set to 45°, on the edge to be mitered, align the rule against the inside face's baseline and draw a line with a pencil at 45° toward the corner to the outside face. If the baseline was marked correctly, these two points should connect.

Starting from the mitered portion's offset from the rear edge, the rest of the board should be treated like through-dovetails. Use whatever layout method you desire. I use dividers to get tail size, starting 1 cm in from the edges when I lay out dovetails. I then square the points on the end grain with a face edge, use a dovetail marker at 1:8 ratio, mark my waste, and cut. The outer edge of the gap between the mitered edge and the tail is cut to 90°, following the marking gauge line set easier. I leave about 1 mm to be chiseled on this corner to make up for potential sloppy sawing.

At this point, the tail board is complete, save for the miter. Between the miter and first tail, there is a gap where the saw may plunge into the tail as it drops if you're not careful when cutting the miter. To prevent this, I place a scrap piece that fits in the gap and secure it downward into my Moxon vise so the scrap rests against the corner to be mitered. With this method, the saw does not drop and only cuts the scrap piece.

If attempting to cut a perfect 45°, this is a good opportunity to execute a first-class saw cut. Instead of a pencil, use a marking knife to scribe the line from the baseline to the corner. Start with a shallow stroke followed by a few progressively harder strokes. With a sharp chisel, pare away the waste area to begin a wall to use as a guide with your saw. Make the wall as deep as you are comfortable with and then saw away the waste.

On the pin board, using the edge to be mitered as a reference, with the same marking gauge settings (for this particular project, the 15 mm setting), score a line on the end grain and the inside face only. Using blue tape for contrast, I transferred the tails and removed the waste.

For the miter on the pin board, take great care when cutting the edge wall against the pin. The saw should tilt at 45° so as to not cut into the front face of the pin board. Once the cut is established, rotate the workpiece 90° in the vice and cut the miter just like the tail board, taking care not to cut into the pin.

To finalize the miter, use a 45° block to assist with the paring. It is a simple piece of scrap cut at 45° with a long-enough edge to support a chisel for paring downward. Aligning the

Protection. I use a scrap piece of wood under the miter (left) as a backer to prevent my saw from cutting into the adjacent tail (right).

Perfect miter. To get an almost perfect miter, use a first-class saw cut to pare away some waste (left) before cutting (right).

guide to the baseline can be tricky. I use my widest plane blade (my No. 7) and lightly set the blade into the marking line. Butt the guide block into the blade and secure it with clamps or holdfasts. Use a freshly sharpened chisel against the guide block and pare down the miter to fit.

HAND-CUT DOVETAILS | Mitered Dovetail Box

Miter on point. I check the miter with a combination square.

Line up the boards. Aligning the tail board to the pins is key. I use the flat back of a chisel against both edges as a reference.

Guide your angle. Use holdfasts or clamps to secure the guide in place; begin paring. This is the most satisfying activity of cutting this joinery.

Mark orientation. Use an arrow to note grain direction for half-blind dovetails. Orient grain to run toward the back of the drawer to allow for easier planing of the pin board end grain when sizing the final fit.

With the miter finalized, cut the groove in the fashion you see fit but be careful with the mitered edge; it is a very delicate area. It is easy to bruise or completely chip the edge. The outer carcase is ready for glue-up.

THE DRAWER

The drawer front is a thicker piece of black walnut perfect for half-blind dovetails. It has a bark inclusion in one of the corners which provides visual appeal. Once the outer carcase finished drying, I sized it to fit perfectly into the carcase, leaving it a little snug to allow for play room later. I milled up a portion of my ash slab and cut the tails into the ash pieces. The contrasting ash with the black walnut made marking the tails on the pin boards easy to follow.

I applied the same technique for the outer carcase back panel for the drawer bottom. This time, however, the raised portion sits inside rather than outward. To simplify the drawer, I used a simple rebate to house the bottom, making sure to not cut all the way through to accidentally expose the rebate in the pins.

When marking baselines for dovetails, it's common to aim for the pins to remain proud. However, for this drawer, I made the tail boards proud of the pins. With the drawer front sized to fit the carcase, it was easier to plane the whole tail board rather than risk accidentally resizing the drawer front. It involved some forethought into the layout, but after cutting the final through-dovetails on the back and finalizing the glue-up, sizing was much easier and the drawer fit like a glove.

Mitered Dovetail Box | HAND-CUT DOVETAILS

FRONT VIEW: CARCASE

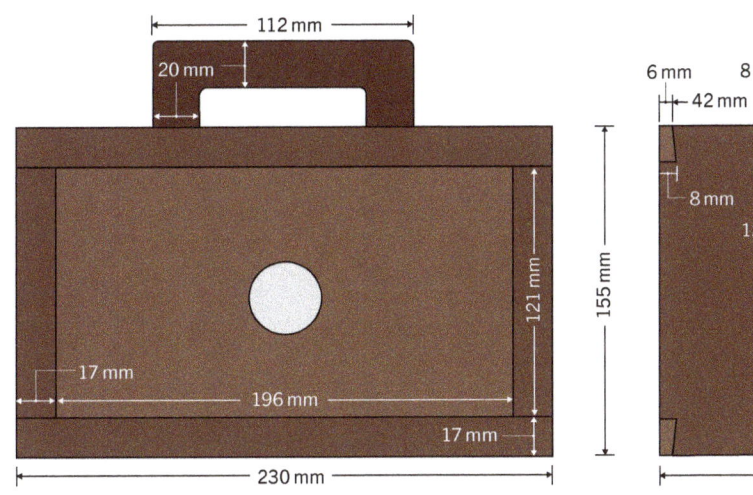

SIDE VIEW: CARCASE

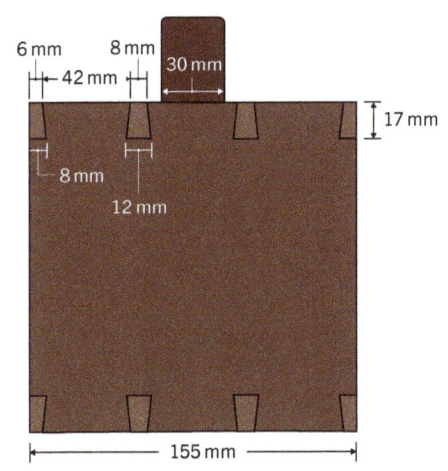

SIDE VIEW: DRAWER

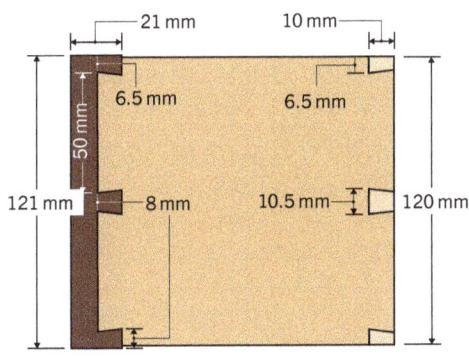

BOTTOM VIEW: DRAWER

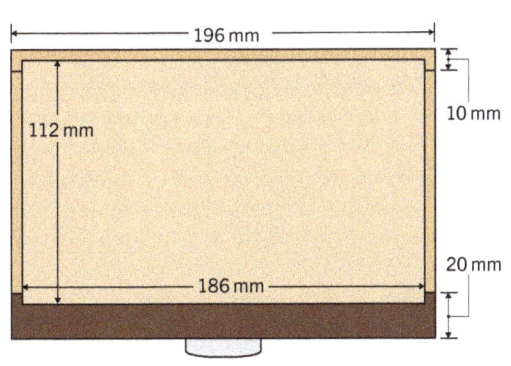

EXPLODED VIEW: CARCASE

3D VIEW: CARCASE

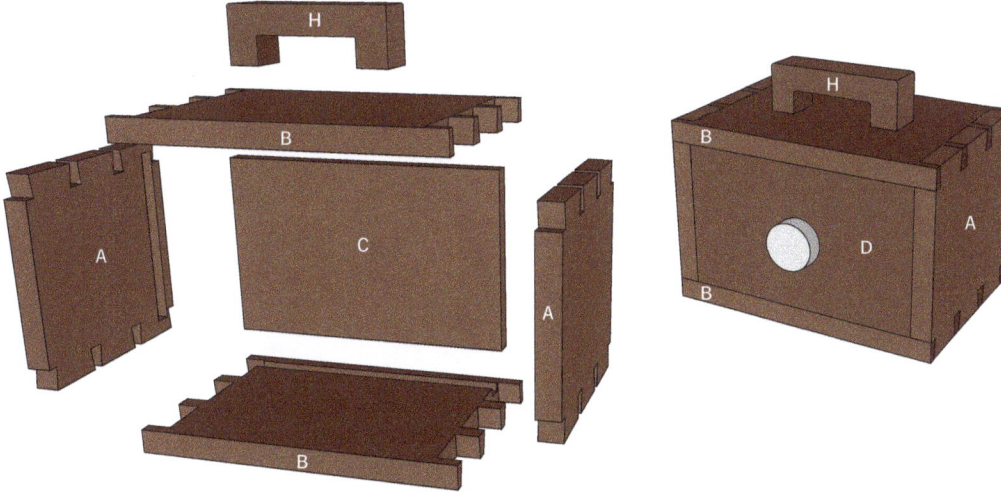

THE DOVETAIL BOOK

HAND-CUT DOVETAILS | Mitered Dovetail Box

Get a handle on it. The handle is shaped for the top of the box.

CUT LIST & MATERIALS

	QTY		ITEM	DIMENSIONS (MM)			MATERIAL
				T	W	L	
☐	2	A	Carcase sides	17	155	155	Black walnut
☐	2	B	Carcase top and bottom	17	155	230	Black walnut
☐	1	C	Carcase back	12	137	212	Black walnut
☐	1	D	Drawer front	21	121	196	Black walnut
☐	1	E	Drawer back	10	120	196	Ash
☐	2	F	Drawer sides	10	112	120	Ash
☐	1	G	Drawer bottom	10	112	186	Ash
☐	1	H	Handle	30	40	112	Black walnut
☐	1	I	Drawer pull				Glass

To convert measurements from mm to inches, divide by 25.4.

FINAL DETAILS AND FINISH

Seeing the box and feeling the weight of it, my wife requested I affix a handle to the top to provide better support for carrying. I had an offcut of black walnut left over that fit the bill. Opportunities to practice with my various shaping tools are rare, so I took my time when designing, roughing out, and shaping the piece. The box was designed with 1:8 ratio dovetails, so I incorporated that ratio into the inside taper of the handle. Slowly and carefully, I used spokeshaves, rasps, files, and sandpaper to round over the edges, added little dimples in the corner, and created something that my wife could hold easily while also being aesthetically pleasing. With a little finesse, I was able to glue and screw the handle into place. This process would have gone more smoothly had it been completed before glue-up, but I managed to make it work.

I chose a simple boiled linseed oil for finish. Yellowing was not a concern since black walnut is dark and the ash would rarely be visible. To apply the boiled linseed oil, I did an initial heavy application after sanding to 600 grit. After about 15 minutes, I wiped away excess oil and let it sit for a full 24 hours. I repeated this step three times over four days and I was left with a very smooth, beautiful finish that still allows for the wood texture to be felt. ■

Mitered Dovetail Box | HAND-CUT DOVETAILS

THE TELEGRAPHING EFFECT

BY MARTIN GRESHOFF

Years ago, dovetails were commonly used to join high-end furniture. It's easy to spot them on the veneered jewelry box at right, isn't it? Over the years, the sides of the box have shrunk in thickness, but not in length. This results in dovetails that are slightly proud of the surface. Their protruding end grain has deformed the veneer on top.

Old-timers had a word for this: telegraphing. In this case, telegraphing doesn't refer to Morse code sent along a wire. It means "sending a message unwittingly," like raising your eyebrows when you're dealt four aces. Only here, you haven't hit the jackpot.

You can find many examples of telegraphing on antique veneered pieces of furniture. Most of these pieces were made using air-dried wood that has since contracted in width and thickness in modern, centrally heated homes. Just about any flaw in the substrate (the wood under the veneer) might show through. Knots, checks, uneven gluelines—all of these eventually come to the surface, so to speak, even though the veneers used were often quite thick.

Does telegraphing rule out the use of dovetails under a veneer? No, you just have to make a modified half-blind joint: a double-lap dovetail or a secret mitered dovetail (which is even trickier to make). When these joints are glued, neither the pins nor the tails show on the outside.

Double-lap dovetails—open.

Double-lap dovetails—closed.

DOVETAIL STATION

This compact stand organizes and stores your favorite tools for making dovetails.

BY TOM CASPAR

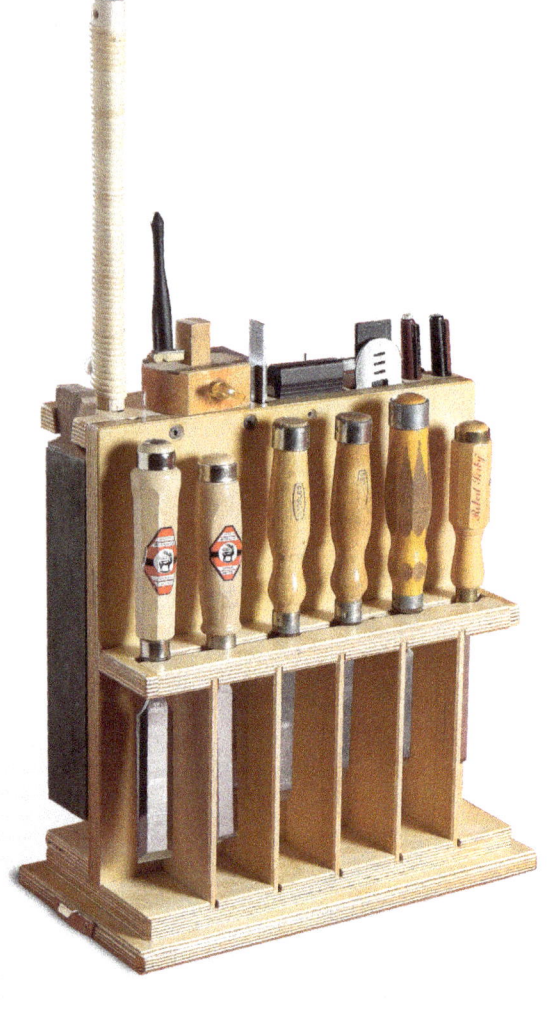

Cutting dovetails by hand is supposed to be an orderly, precise process. But when you're in the thick of it, it's all too easy to end up with a bench littered with stray tools and precious little free space left over for the actual work.

I built this station to keep all my tools in one place, neatly organized and out of the way. When I'm temporarily done with a tool, I place it right back in its home. I can reach most tools from the front of the station—the chisels are in front (above right), while the layout tools are on top.

When I'm done with my work, I take apart my saw and store it in back of the station (above left)—along with some other tools—then slide and lock a cover over the station.

I'll admit it's a bit extravagant to dedicate a set of tools just for dovetailing, since some can be used for many other jobs, but I think it makes practical sense. When I'm ready to chop, the tools are ready, too.

Dovetail Station | **HAND-CUT DOVETAILS**

Tools are close at hand. The station has a convenient home for every tool you'll need.

It's a storage box, too. Slide a cover over the station to protect your tools when you stow them away.

Secure latches. Sturdy window latches lock the cover in place for transport.

BUILD TWO TOOL RACKS

Before you begin making this station, take stock of the tools you'll use for dovetailing. I designed the station to hold my favorites, but you'll probably need to alter the station's dimensions or configuration to accommodate your set. (See "Tools for Dovetailing," p. 6, for a closer look at the tools I designed my station to hold.) Figuring out the most compact arrangement of your tools will take some experimenting. Here's how to go about it.

Start with the layout-tool rack. Make this from scrapwood first, so you can modify it as you go or make a fresh start if a completely different arrangement proves to be necessary. Lay out a series of notches or holes to hold your tools, then cut the notches on the bandsaw.

Cut a center panel—also from scrapwood—and fasten the rack to the panel. Clamp the panel in a vise with the rack facing away from you. Place all of your tools in the rack and see if there's enough space around each one. Ideally, you should be able to remove or replace each tool without disturbing its neighbors. In addition, no tool should stick out beyond the edges of the rack. (If any tools stick out, the station's cover won't fit.)

I had to take apart my trial rack and panel a few times in order to arrive at an optimal arrangement. Along the way, I found that I had to saw a short groove in the center panel to accommodate the blade of my small double square. I also sawed a long groove to hold the blade of my dozuki saw (it comes apart for storage; see p. 97, Back View). And I had to deeply countersink the hole for my striking knife—and add a filler—so the knife would stand up straight. If you have any tools that need to hang on the back of the station for storage, such as a dozuki saw's handle, now's the time to work out their positions. When you're done, every tool should have a secure home. Here's the acid test: No tool should fall off when you move the station.

THE DOVETAIL BOOK | 91

HAND-CUT DOVETAILS | Dovetail Station

Begin building the station. Design a rack to fit your layout and cutting tools. Use a bandsaw and miter gauge to cut notches for the tools.

Baltic birch plywood

Round over the rack's edges. A solid-pilot bit is ideal for getting into narrow cutouts—the pilot is only 3/16" diameter.

Solid pivot

Make a second rack. Drill and countersink holes sized to fit the chisel sockets.

Once you've got your own design worked out, make the actual rack and center panel from Baltic birch plywood. (All the other parts of the station are made from Baltic birch, too.) Round over the exposed edges of the rack on both top and bottom sides (middle photo). A solid-piloted router bit works best to get into tight spots (I used a ¼" radius roundover carbide cutter, 3/16"-diameter solid brass pilot from MLCS).

Next, design the chisel rack. I have a large set of chisels to choose from, depending on the size and type of dovetails I'll be working on, but a compact box can't hold all of them. That's OK, because I don't need the whole set for most projects—just a few will do. After making my layout-tool rack, I found that the station's width would accommodate six chisels spaced comfortably apart. For most jobs, that's just about right.

I often use wide chisels when dovetailing, and this rack is designed to hold them in any location. To build the rack, drill and countersink holes in a single piece of plywood (bottom photo). Countersinking isn't required, but I found that my chisels stood upright more easily if the holes had deeply beveled sides. You'll need a countersink with an extra-large

STRAIGHT BANDSAW CUTS

Use a miter gauge to make sure your bandsaw cuts are straight and square. If you attach a long fence to it, you can support work on both sides of the blade (see top photo).

Dovetail Station | **HAND-CUT DOVETAILS**

diameter (I used one with 82°). Rip the piece of plywood 3/16" off the center of the holes (top). Make two end caps for the rack and a couple of 3/8"-thick spacers, then glue the rack together (center). Round over its exposed edges, top, and bottom.

Make the station's base, then cut 1/4" x 1/4" dadoes in the chisel rack and base to accommodate dividers for protecting the chisels' edges (bottom; p. 96). Make the dividers from solid wood to fit tight in the dadoes.

Leaving the 1/4" dado set in the saw, cut a groove in the base to hold the center panel (see p. 96).

Here's how to do it: Start by sawing a groove slightly off center, then spin the base around and saw a second groove. Shift the fence about 1/8" farther away from the blade and repeat the procedure (p. 94, top left). Continue in this manner to widen the groove, moving the fence a bit less each time, until the groove fits the panel.

Rip the chisel rack. It should be 3/16" off the center of the holes. Make this cut from both sides.

Place 3/8" spacers. They should go between the pieces. Glue caps on both ends. The resulting slot between the holes accommodates wide chisels.

Cut dadoes in the stand's base. Also dado the chisel rack to house dividers. Clamp a block to your saw to limit the length of the dadoes.

STRONG DADO JOINERY

To make a strong joint when using plywood, a groove or dado should be just wide enough for a tight fit. Most plywood is thinner than its nominal size, so the groove or dado must be undersized, too. To make an undersized cut with a table saw, you can either add shims to your dado set or make multiple passes. The second option is faster for making just a single dado. It also allows you to center the groove or dado in the middle of your stock, which is what's required here.

THE DOVETAIL BOOK | 93

HAND-CUT DOVETAILS | Dovetail Station

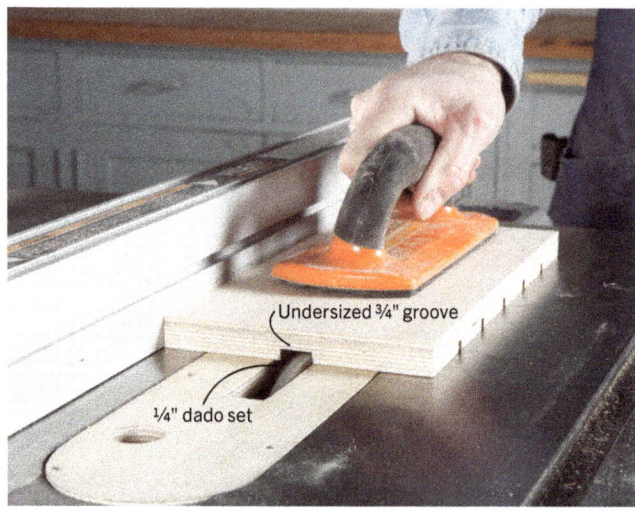

Cut a groove in the base to hold a vertical panel. For a tight fit, make multiple passes from both sides of the base, slightly moving the fence each time to widen the groove.

Fasten the base to the panel. Don't glue it—you may want to disassemble and modify the station later on to store new tools.

Assemble. Fasten the layout-tool and chisel racks to the panel with screws. Place dividers under the chisel rack to hold it in place.

Begin making the station's cover. Rout miters on each of its pieces. The miters shouldn't come to a point—leave a tiny blunt edge.

Fasten the base to the panel (top right). Make sure the screw holes are offset from the dadoes in the base. Fasten the layout-tool rack to the panel (p. 96, Front View). Place the dividers in position, then fasten the chisel rack to the base (bottom left).

I made a few additional pieces to hold my tools and nailed them in place with short brads. I also hung some tools on brad nails. After hammering in the nails, I bent them slightly upright with a pair of pliers as insurance that the tools wouldn't slide off when I moved the station.

MAKE THE COVER

Saw the cover's bottom, top, front, back, and sides to final size. Check these dimensions before you saw, of

Dovetail Station | HAND-CUT DOVETAILS

Tape the joints. Place the pieces against a straightedge, face up, and join them with tape. Use tape that can stretch a bit.

Apply glue. Turn over the pieces and apply glue to the miters.

Fold the pieces into a box. Add band clamps to ensure that the joints come tight.

Glue the cover to the box. The cover's edges are mitered, too. Once the glue is dry, rout a large chamfer around all of the cover's edges.

course. Allow 1/16" clearance between the inside of the cover and the outside of the station's base, all around. In addition, make sure the cover is tall enough to accommodate all your tools when they're stowed away.

Rout miters on the front, back, sides, and top pieces using a chamfer bit (bottom right). Take at least two passes, removing a little more each time. To guarantee that the panels stay full size (since their dimensions are critical), leave a 1/64" wide blunt edge on the last pass. After you're done routing, run your fingers over each miter to make sure it's smooth and free from lumps. This will ensure that your joints fit tight.

Gluing together a mitered box isn't difficult; usually, you don't need

HAND-CUT DOVETAILS | Dovetail Station

3D VIEW: COVER

3D VIEW: STATION FRONT

3D VIEW: STATION BACK

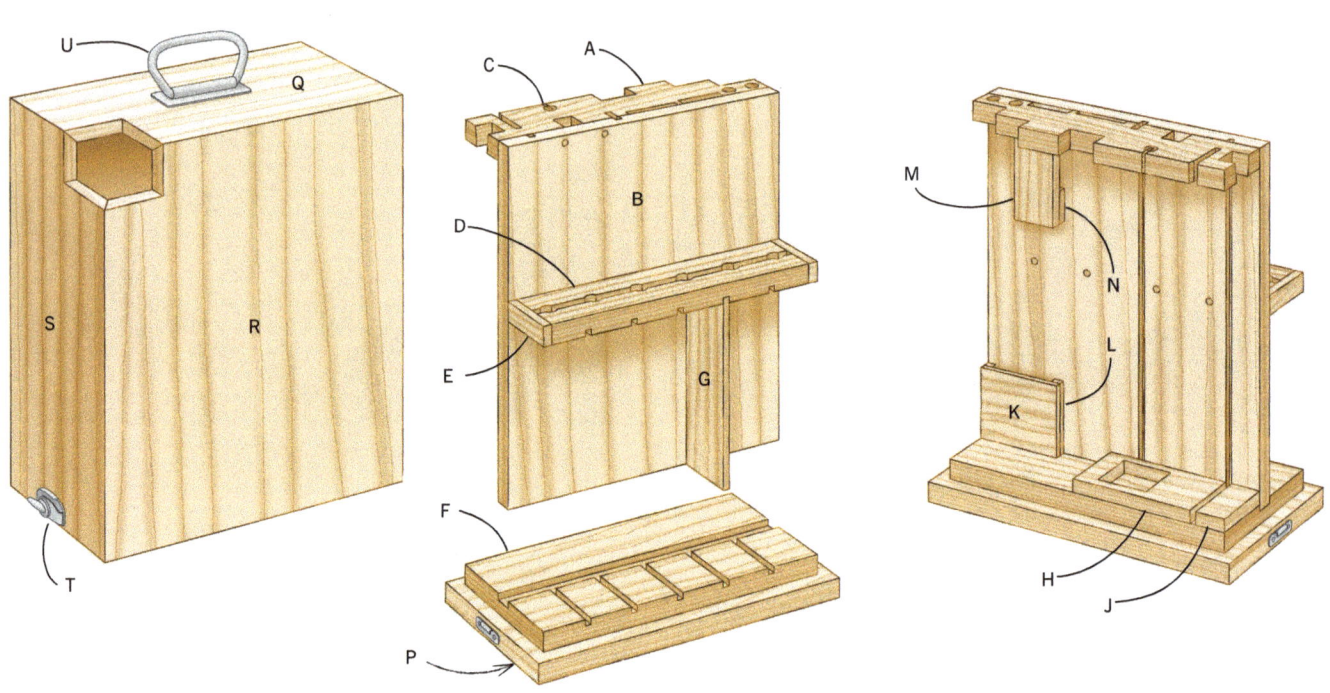

FRONT VIEW

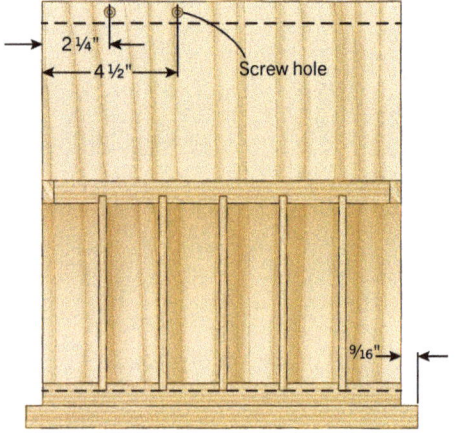

LAYOUT, TOP VIEW: LAYOUT-TOOL RACK, CENTER PANEL, CHISEL RACK, AND END CAPS

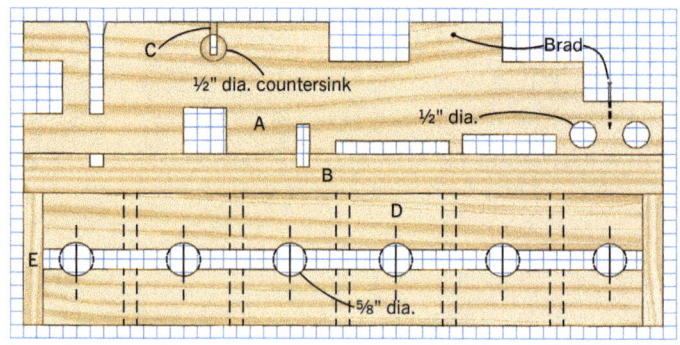

Scale: 1 square = ¼"

Dovetail Station | **HAND-CUT DOVETAILS**

TOP VIEW: BASE

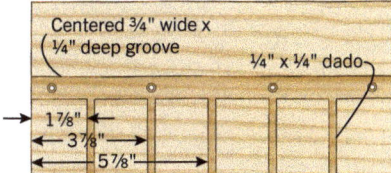

BACK VIEW

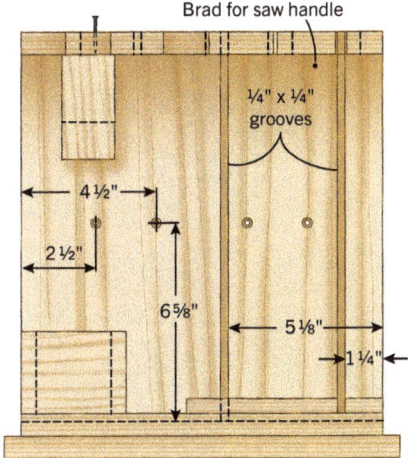

CUT LIST & MATERIALS						
	QTY		ITEM	DIMENSIONS (INCHES)		
				T	W	L
STAND						
☐	1	A	Layout-tool rack*	¾	2 ½	12
☐	1	B	Center panel	¾	12	13
☐	1	C	Filler	1/16	5/16	¾
☐	2	D	Chisel rack	¾	1 1/16	11 ¼†
☐	2	E	End cap	¾	⅜	2 ½
☐	1	F	Base	¾	5 ⅞	12
☐	5	G	Divider	¼	2 ½	6 ½
☐	1	H	Mallet block	½	2 ½	5
☐	1	J	Saw blade holder	½	2 ½	1 ¼
☐	1	K	Shim pocket 1	¼	3 ½	2 ¾
☐	2	L	Shim pocket 2	¼	½	2 ¾
☐	1	M	Thin blade 1	½	1 ¾	3 ½
☐	1	N	Thin blade 2	½	1 ¾	1 ¼
COVER						
☐	1	P	Bottom	¾	7	13 ⅛
☐	1	Q	Top	½	7	13 ⅛
☐	2	R	Front and back	½	13 ⅛	17
☐	2	S	Side	½	7	17
SUPPLIES						
☐	2	T	Sash lock			
☐	1	U	Handle			3 ½

Overall dimensions: 17 ¾" H x 13 ⅛" W x 7" D.
Note: * All wooden parts are Baltic birch plywood. † Start with a single piece measuring ¾" x 2 ½" x 11 ¼".

clamps, just tape. Using the "blunt edge" strategy, however, the tape must have a little bit of stretch in it. I used 1"-wide reinforced duct tape, such as Gorilla tape. It stuck well, with just enough stretch, but didn't yank out the wood's fibers when I removed it.

To prepare the cover for gluing, arrange the front, back, and sides in order, flat on your work surface, face up. Butt the pieces tight together and align them with a straightedge. Join the pieces with 6" long pieces of tape (p. 95, top left). Turn over the assembly and apply glue (p. 95, top right).

Fold the pieces together and tape the remaining sides (p. 95, bottom left). Tape alone should keep the joints tight, but I added a couple of band clamps to make sure. Glue and clamp the top (p. 95, bottom right). Rout a 3/16"-wide chamfer all the way around the cover—this will remove the blunt edges of the miters. Add two sash locks (one on each side) and a handle. You're all set! ∎

SHAKER TRAY WITH A LITTLE EMBELLISHMENT

Requiring sawing, planing, and shaping skills, this tray is the perfect project to put hand-tool lessons to practice.

BY KENAN ORHAN

This tray is a slightly amended version of a shaker cutlery tray and is a great one-day project to help you hone your hand-tool skills while still producing a beautiful and utilitarian piece. I start by breaking down rough cherry lumber and mill pieces to thickness and length. Though the divider that makes up the handle will be slightly shorter than the long walls of the tray, I cut

Shaker Tray with a Little Embellishment | HAND-CUT DOVETAILS

Lay out the tails on the front and the back of the tray.

Angle the tail boards in the vise. You should be sawing straight down.

Remove the waste from between the dovetails.

Use a chisel to clean up any stray fibers. The corners will need help.

it the same length and will adjust it later. With planning and care, you can create an effect with the grain that wraps around from one face to the next, but since this is dovetail construction instead of miters, I don't bother too much with that, preferring just to lay out the faces so the grain is pleasing to the eye.

GANG-CUT TAILS

I start my dovetails by gang-cutting the tails. This method speeds up the process considerably. The trick with gang-cutting, however, is really paying attention to the baseline so that you don't extend your kerf beyond it. Having a dead-level cutting action will help, but even I struggle with this sometimes, so as I near the line, I check the opposite face as well, sort of teetering between them. Then when the kerf touches the baseline on both faces, I use a very light touch and a few passes to make sure the kerf goes straight across rather than leaving a bump of unsawn waste in the middle.

HAND-CUT DOVETAILS | Shaker Tray with a Little Embellishment

Transfer the tails to the pin boards.

Cut the pins. Stay right on the edge of the knife mark.

Grab a float or file. This is a great way to fine-tune a joint and keep the pins straight.

While cutting your tails, something to consider if you find that your line drifts, is to position your boards so that the line you follow with your saw is perpendicular to the floor. This really lets gravity assist with that straight line, and considering the secret to good dovetails is straight lines, that's an assist that's hard to pass up.

TRANSFER TAILS TO PINS

When it comes time to transfer tails to pins, there are a lot of ways to get the boards lined up just so, from using special jigs and offsets to just holding it there with a hand and support. I've found the best way for me to get a square transfer right up to the baseline is to add a piece of blue tape to the tail board just under the baseline. This will register with the corner of the pin board much in the same way others use a little rabbet on the inside of their tails, but I don't have to use a plane to do it or worry about blow-out. Then, make simple knife marks with light, successive scoring to avoid forcing an incorrect line.

When I go to cut the waste, I slip the very outside tip of the saw tooth into that knife line so I won't have to pare any waste away. I've found that when paring pins, more often than not, the grain doesn't cooperate with me and pulls my chisel too deep into the cut, removing part of the wood I wanted to keep. When you cut right at the edge of your knife line, you mostly remove this need to shave material, and your boards should come together off the saw. If they're still a little tight, I recommend a file to take very light passes, maybe three at a time, off the tip spots until you

Shaker Tray with a Little Embellishment | HAND-CUT DOVETAILS

get a snug fit. If, however, you don't trust yourself to get that close or you know that you drift, go ahead and give yourself a half or full saw kerf of extra space into the marked waste, especially if you're working with nice wood, but you should know that the only way to get to the point of sawing right on the line is by doing it in practice over and over again!

NEXT UP, THE HANDLE

With the tray walls ready, it's time for the handle. I take the inside distance directly from the long wall, from scribe line to scribe line.

This establishes the length of the handle minus its tongues, each of which will be ¼", so the board will be the interior length plus ½". When it's cut, I make sure everything is square at a shooting board, then transfer tick marks from the long wall again to mark where I will have to cut the notches for the handle board. Taking measurements from the pieces you have already made is the best way to guarantee dimensions match in crucial moments. I always recommend avoiding measuring with tapes and squares if you can avoid it.

Since I've made a lot of these trays, I use a stencil to lay out the curve and the finger slot. You can come up with your own rather easily by cutting out one half of the pattern from card stock or thin MDF and flipping it to make both halves so that you have a symmetrical handle, or if you're a free spirit, freehand it! Either way, trust your eyes, as they are good judges of symmetry. With it ready, I cut the line with a coping saw and smooth the cut with two spokeshaves: the small, round bottom one

The divider gets a swooping cut. Do this on the edges with a coping saw.

Refine the divider edges. Use a spokeshave.

Smooth the shape. After drilling out the handle, use a file or rasp to smooth.

HAND-CUT DOVETAILS | Shaker Tray with a Little Embellishment

Chop a dado in the ends for the divider.

Chisel out the dado waste.

Bring the dado to a consistent depth. The router plane is the perfect tool for cleaning up the bottom of a dado or groove.

Prep for gluing. Use a handplane to smooth out the faces on all of the parts.

for the concave and a flat bottom one for the convex. Take your time here; spokeshaves are wonderful tools, but they work best removing nice thin shavings, leaving a glass-smooth surface. If you want, you can cut at the band saw, but you'll still need to smooth that rough surface.

If you want to stick with hand tools for the finger slot, use a small bit to drill a hole near the line. Fit a coping saw blade through it and cut the hole. Alternatively, you can use a wide bit and a brace to remove a lot of the material, then file it smooth with an aggressive rasp. I did the latter, but because my tool was acting up, I used a Forstner bit at the drill press to remove large chunks of waste, then rasped the ridges between the cuts flat and filed them smooth.

With the handle cut, remove the waste at either end of the board that creates the tongue that will fit into the stopped dado we will make in the short walls. I use a chisel to chop right to my scribe line for a good, snug fit.

Shaker Tray with a Little Embellishment | HAND-CUT DOVETAILS

It's time for the dado in the short wall. I measure out to the middle of the board, then make a mark ¼" off center (half the thickness of the handle). Using the handle board, I register it to the tick and make another tick on the other side of the halfway point so that I have the dadoes' wall measured exactly to the thickness of the board. Next, I use a chisel in much the same way a blind mortise is made — establishing the dado walls before using a router plane to remove the waste down to the appropriate depth (¼"). Test the fit of everything by putting the handle into its dadoes, then tapping the tail boards lightly into the pin boards, but don't seat them! I don't even press them in; just get them started and registered to make sure the handle board doesn't prevent them from coming together. If the handle board is a little long, take it to the shooting board and swipe a few shavings off, and test again.

PRE-FINISH FIRST

I like to pre-finish the inside before glue up. This ensures my interior has a perfectly applied finish but also allows for easier removal of glue squeeze-out. For most of my projects, especially small, tactile pieces, I use mineral oil followed by a coat or two of furniture wax as a finish. This gives it a nice, muted luster and leaves the wood feeling like wood instead of plastic. I glue the tray up and slide in the middle section.

THEN ASSEMBLE

Once the tray is dry, I plane the tails and pins flush and use a pin gun with a small bead of glue at the side to attach the base. At this size and

Carefully glue up the tray. A little bit of glue on the inside of the pins is all that's needed for a rock-solid joint.

Install the divider. Slip it into the dado from the bottom of the tray.

Clamp. Allow the glue to cure. F-style clamps allow pressure to be applied directly to the tails.

HAND-CUT DOVETAILS | Shaker Tray with a Little Embellishment

Attach the bottom. After cutting the bottom, nail it in place.

Start the wraps. Hold the tail against the handle. Then, as you start wrapping, pinch the tail under the wraps to hold it in place.

CUT LIST & MATERIALS

	QTY		ITEM	DIMENSIONS (INCHES)			MATERIAL
				T	W	L	
☐	2	A	Long sides	½	3 ½	14	Cherry
☐	2	B	Short sides	½	3 ½	8	Cherry
☐	1	C	Divider/handle	½	5 ½	13 ½	Cherry
☐	1	D	Bottom	¼	8	14	Cherry
☐	1		Cord wrap				Hemp cord

with the thickness of the cherry, the wood movement isn't enough of a factor to give me concern. With the bottom attached, I plane it flush and add a slight chamfer along all outside edges. Then I apply the rest of my finish before moving on to the cord wrapping.

ADD THE CORD WRAP

Wrapping the cord can feel a little tedious, but it should be quick work at this scale. I use hemp cord from my local hobby shop and start by securing 2" of cord from the end to the underside of the handle. This can be done with glue, tape, or a pin, but I find it's easiest to just hold the end on the bottom of the handle to one side and then wrap it so that the cord wraps itself over the end and secures it in place. Then it's just a matter of wrapping until you near the other end of the handle. With maybe 2" of wrapping left, I put a bent paperclip under the handle and wrap the cord around it much like I did with the start, but when I'm finished wrapping, I slip the remainder of the cord through the paperclip and pull it out, so that the cord is securing itself at both ends. Then I give the tail end a little tug to snug it down and cut the cord. Now, the tray's ready to be put to use in your home. ■

Shaker Tray with a Little Embellishment | HAND-CUT DOVETAILS

Wrap a paperclip. When you get to the end, wrap a paperclip under five or six wraps. Feed the free end through the paperclip eye and pull the tail under those final wraps.

FRONT VIEW: DIVIDER/HANDLE

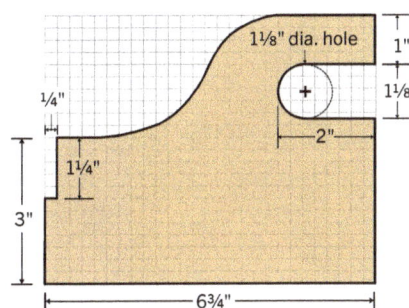

1 grid square = ¼" square

Complete the wrapping. Once the tail is through the wraps, grab it and pull it tight. You may have to "tidy up" the wraps, but once they're in place, they'll stay. Trim the end flush.

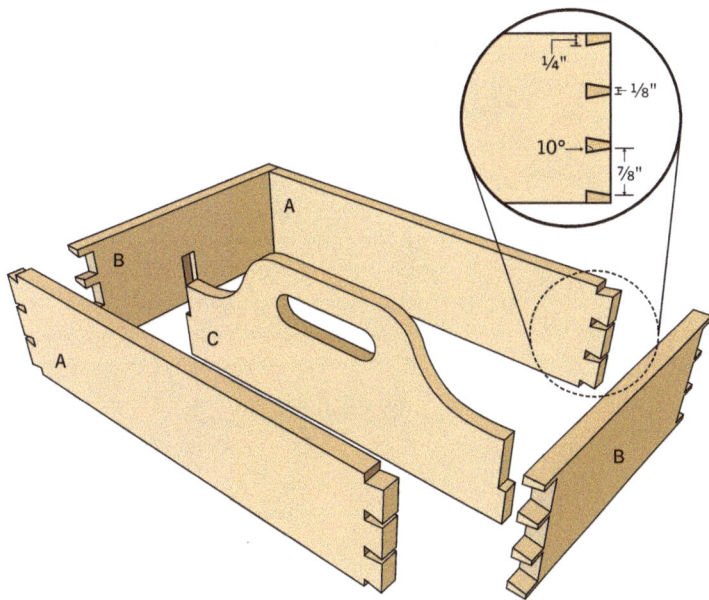

THE DOVETAIL BOOK | 105

2

POWER-CUT DOVETAILS

On some occasions, you just don't have time to hand-cut every dovetail. In this section, you'll find advice from power-tool pros on using your table saw, handheld router, router table, and even your band saw to cut perfect dovetail joints. Traditional, half-blind, and sliding dovetails are covered—even giant-size condor tails for your workbench. If you've been avoiding your dovetail router jig, you'll want to visit the articles near the end of the section that reveal all the tips and tricks you need to master the jig. And for those who want to speed up the process but still have a hand-crafted touch, check out Scott Gibson's article on combining power tools with hand tools (p. 113).

BANDSAWN DOVETAILS

Get a hand-cut appearance with half the fuss.

BY SETH KELLER

If you've labored over hand-cut through-dovetails, you'll be amazed how much faster they can be cut on the bandsaw. You get all the benefits, including strong joints, classic appearance, the ability to use boards of any thickness, and the freedom to size and space the pins and tails however you want. The only limiting factor is your bandsaw's throat capacity. My saw allows making joints up to 14" wide. That's wide enough for any drawer, but not for a blanket chest.

As with any technique, mastering this one takes a little practice. You'll need a sharp blade for your bandsaw. I keep a chisel handy, too, for fine-tuning the fit.

Bandsawn Dovetails | POWER-CUT DOVETAILS

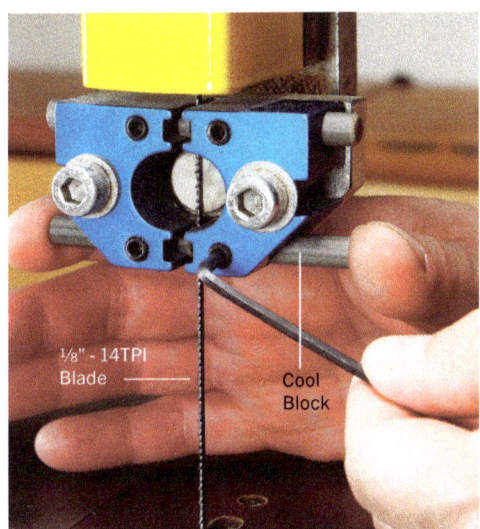

Outfit your bandsaw. Mount a ⅛" blade with 14 teeth per inch (tpi). Bandsawn dovetails require making sharp turns in confined spaces. You'll also have to replace your saw's metal or ceramic guide blocks with Cool Blocks. Cool Blocks support the thin blade without damaging its teeth.

Sleds that modify sawing angle. An angled sled allows for cutting the pins without tilting the bandsaw's table. Sleds with different angles cut pins that slope differently. Here, I'll use the 10° sled. It cuts pins that slope at a ratio of about 1:6.

MAKE THE JIG

The jig is an angled sled. The slope angle you've chosen for the pins determines the sled's angle. Ratios of 1:6 and 1:8 are commonly used to determine pin slope angles. To create pins with a 1:6 slope, make a sled that slopes at 10°.

Cut the angled sides with your miter saw—you can cut both sides at once by centering a wide piece and making one cut. You could also cut the angled sides on the bandsaw or table saw, using the fence and a tapering jig. Make sure the angled pieces are identical. Then simply glue and nail the parts together to create the angled sled.

This jig makes it easy to cut both sides of the pins. After cutting all sides that slope in one direction, you simply rotate the sled and reposition the workpiece to cut the sides that slope the other direction.

Locate the pins. Mark on the end of one board. Then use an adjustable square to transfer the straight lines to the remaining pin board faces. Layout of bandsawn dovetails is easier than hand-cut, because you don't have to mark the wedge-shaped pins on the end of every board.

POWER-CUT DOVETAILS | **Bandsawn Dovetails**

LAY OUT AND CUT PINS

Lay out the pins on the end of one board, after scribing the board thickness onto the ends of all the boards (p. 109, bottom). As with hand-cut dovetails, the number of pins, their spacing, and the angle at which they slope is up to you. Typically, half pins are used at both ends of the board.

Before you start cutting, strike lines on the end of the board to indicate the cutting angle. These lines aren't precise, they're simply indicators. Use them to make sure the sled is oriented correctly and to assure you cut on the correct lines.

Place the workpiece on the sled against its fence and make a straight cut to the scribe line on one side of each pin (top). I make these cuts freehand, but you could also use the bandsaw's fence.

Go back and widen each saw kerf (center). To cut the other side of the pins, keep the workpiece facing the same direction, but rotate your sled 180°, so it slopes in the opposite direction (bottom).

To clean out the waste and establish straight shoulders between the pins, remove the workpiece from the sled and flip it over. First, cut an arc to the scribe mark between each pin (p. 111, top left). Do not cut beyond the scribe line. Rotate the board to cut the shoulder (p. 111, top right). Set the fence so the blade cuts precisely at the scribe line.

Check the shoulders you've just cut to make sure they're straight and smooth. Use a chisel to pare any rough edges that remain from the straight-in cuts you made to widen the angled kerfs.

Cut the first side of all the pins. Make the straight cuts following your pencil lines. Stop at the board-thickness scribe mark. Your angled sled automatically slopes the pin. I strike angled marks on the ends to avoid cutting the wrong lines.

Widen each kerf. Make multiple adjacent passes. Always stop the cut at the scribe line. This step makes it easier to cut the shoulders.

Other side of the pin. Rotate the sled, reposition the pin board, and cut the other side of each pin. Again, always stop at the scribe line. Widen these cuts as well, by making adjacent cuts.

Bandsawn Dovetails | POWER-CUT DOVETAILS

Remove pin waste. Flip the board over and lay it flat. Flipping the board allows cutting between the pins' wide ends. Start by making a curved cut through the waste area to the back corner of each pin. End each cut precisely at the scribe line.

Shoulders. Rotate the pin board to cut the shoulders. Using the fence guarantees straight cuts, which are necessary for the joints to fit properly.

Tail board layout. Transfer the pin locations to the tail boards. Hold the pin board flush with the edges and end of the tail board and mark with a fine pencil. Hold the lead tight against each pin, so your lines exactly outline the pins.

CUT THE TAILS AND TEST FIT

When the pin boards are complete, transfer the pins to the tail board (left). Define the pin sockets by making angled cuts to the scribe line. Once the sockets are defined, nibble out the waste and cut the half pin shoulders. Work slowly, being sure to never go over the line. Press the pieces together. Ideally, they'll slip together with light pressure. If you have to use a mallet, they're too tight: When you apply glue, they won't go at all. ■

POWER-CUT DOVETAILS | Bandsawn Dovetails

Cut. Cutting the pin sockets creates the tails. Define the sockets by cutting along the inside edges of the pencil lines. Be sure to leave the lines. Remember, they outline the pins.

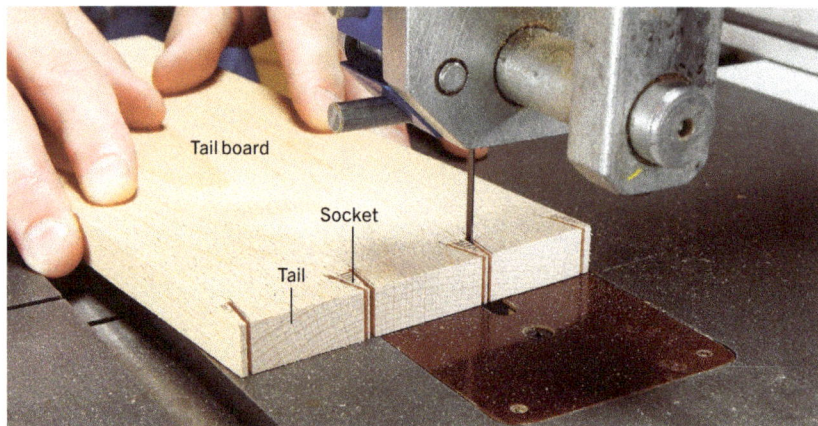

Complete the pin sockets. Nibble out the waste between two angled cuts. Rotate the workpiece 10° to cut the shoulders of the half pins.

Test-fit the joint. The parts should slide together with light pressure. If they won't go, locate the spots that bind and pare them to fit with a chisel.

POWER-ASSISTED HALF-BLIND DOVETAILS

A small router takes the drudgery out of half-blind dovetails.

BY SCOTT GIBSON

It's hard to blame anyone for turning to a router and template when a job calls for a kitchen's worth of dovetailed drawers. If nothing else, the process is fast. Once you've fine-tuned the router settings, you can knock out a lot of drawer boxes quickly and with perfect regularity.

POWER-CUT DOVETAILS | Power-Assisted Half-Blind Dovetails

Look-alike. Half-blind dovetails cut using this method look exactly like those cut entirely by hand.

First base line. Using a marking gauge, scribe a line across the end grain of the drawer front. It should be about one-third of the board's thickness in from the face.

Second base line. Without changing the setting on the gauge, scribe a line on both inside and outside faces of the drawer sides. These are the base lines for cutting the tails.

and slope are fixed by the shape of the router bit, and bits may have thicker profiles than you would like.

These characteristics aren't as important when the drawers are going into a kitchen. But when a drawer is bound for something special—a chest of drawers, for example, or a nice little side table—the look and lack of flexibility in a machine-cut dovetail is all wrong. Hand-cut dovetails have a lot more eye appeal, and their dimensions and spacing can be infinitely adjusted to suit different tastes, species of wood, and drawer heights. The only problem is that cutting them by hand and removing all the waste with a chisel and mallet is tedious—about as interesting as watering the lawn.

One way to speed up the process without detracting in any way from the hand-cut look is to use a small router to do a lot of the heavy stock removal. I use a laminate trimmer, a small router that can be guided easily with one hand. This approach takes half the time of doing all the work by hand, and with a little practice produces consistently accurate results. Best of all, you won't be limited by a template or the shape of a router bit. It's really the best of both worlds.

LAY OUT AND CUT THE TAILS FIRST

Much of the process is identical to doing it all by hand; the first thing is to prepare the drawer parts carefully. Assuming the drawer is to fit in a case, without the use of any drawer hardware, an accurate fit is essential if the drawer is to work smoothly.

It may seem counter-intuitive, but making a drawer loose for the opening makes it harder to operate

And that's the problem—all that regularity makes for boring joinery.

Boring and uniform: With many dovetail jigs, the spacing of tails and pins is fixed, and those pin-to-pin dimensions don't always coincide with actual drawer heights. Templates that allow adjustable spacing are an improvement, but pin width

Power-Assisted Half-Blind Dovetails | **POWER-CUT DOVETAILS**

Reset the gauge. It should equal the thickness of the drawer side and scribe a line across the inside face of the drawer front. This represents the depth of the dovetail pins.

Save time. Using a chisel to mark out the tails saves time. Just align the center of the chisel on a mark representing the center of each tail and mark the edges. Then use the bevel to draw the tail locations.

Layout. Dovetail spacing is completely flexible, although the layout usually begins and ends with a half pin. You can make a small *X* where waste is to be removed.

because the drawer racks and binds as it's opened and closed.

The drawer front should just fit into the opening side-to-side. Be sure to allow enough clearance in height so that seasonal changes in moisture content won't swell the drawer closed in summer heat and humidity. The drawer sides and the drawer front must be cut dead-on square for everything to work correctly.

With stock prepared, lay out the joint just as you would if no machine were involved. I start by using a marking gauge to strike a line on both ends of the drawer front about a third of the drawer's thickness in from the front. Without changing the setting, strike a line across each drawer side. This represents the length of the dovetails, and while both dimensions must be equal, the exact length isn't important. Leaving a narrower margin at the front of the drawer looks a little more graceful to me, but it also increases the risk

slightly that something will go awry later in the process.

With these lines scribed, change the setting on the gauge so it equals the thickness of the drawer side and mark the inside surface of the drawer front. This represents the length of the pins and I try to get it as close to the actual drawer side thickness as possible: In case work, it's more common to lay out the joint so that the pins will protrude slightly when the piece is assembled. They can be belt-sanded or planed flush with case sides very easily. But taking too much stock off a drawer side after the joint has been glued up will ruin the fit.

With lines scribed, set an adjustable bevel for the angle of the tails. By convention, the slope on the sides of the dovetail is 1:6 for softwoods and 1:8 in hardwoods, although I'm not sure it matters that much.

Lay out the tail spacing of the tails along the drawer side with a pencil. Spacing is entirely up to you,

THE DOVETAIL BOOK | 115

POWER-CUT DOVETAILS | Power-Assisted Half-Blind Dovetails

Begin to cut. Cut out the tails on the drawer sides with a dovetail saw, taking care not to go past the scribed base line. Unless the cuts are square across the edge of the board, it will be impossible to transfer your pin locations accurately to the drawer front.

Chop out the waste. With the sides of the tails defined by the saw kerfs, use the chisel to chop out the waste. It helps to make the scribe line a little deeper with a knife and pare a small shoulder with a chisel. This prevents the chisel from drifting over the scribe line as you chop.

Chop. Use a mallet to chop downward at the base line, then pop out a sliver of waste by working in from the edge of the board. Take it a little bit at a time and work about halfway through the board. Then flip the board over and finish from the other side.

Clean out. Check that the tails are clean all the way through, with no remaining waste in the corners. It's also a good idea to check that the cuts are square across the end of the board—if not, correct the problem now by paring the tails with a chisel.

although it's typical to start and end the layout with a half pin. It also saves time to use a chisel to set the width of the tail at its widest point.

Now cut out the tails on the drawer sides with a saw all the way to (but not beyond) the scribed base line. Split the line to the waste side, as furniture maker and author Tage Frid used to say, and you won't go wrong.

The half-pin waste on each end can be removed with just a saw. The waste between tails can be chopped out with a chisel. Work from both sides, and when you're finished, make sure all the waste has been removed from the inside corners of the tails. Use a chisel or knife to clean out what's left.

Because tails are cut first, it doesn't matter that much if the angle of the cut is slightly off, or if one tail is slightly wider than another. What does matter, however, is getting the saw cuts square to the end of the drawer sides. If those cuts are angled, it won't be possible to transfer the pin layouts accurately to the drawer front, and the joint won't fit properly. Inspect the cuts carefully, and if they're not square, true them up with a chisel before going ahead.

USE TAILS TO LAY OUT PINS

The completed tails are used to lay out the pins on the drawer front. I put the drawer front in a bench vise and arrange the drawer side so the front edge lines up exactly on the scribed line on the drawer front. Hold the drawer side firmly and use a sharp layout knife to trace the pins on the end of the drawer front. Make sure the drawer side doesn't shift.

This would normally be the time to cut to the lines with a saw and chop out the waste with a chisel; here's where you'll bless the person who invented the laminate trimmer.

What makes it ideal for half-blind dovetails is its small size and light weight, and the fact that it can be guided with one hand. I use a ¼" solid carbide spiral up-cut bit, which makes a very clean cut.

When all the lines have been transferred to the drawer front, clamp a piece of straight wood to the front side of the drawer front to give the router base a wider contact surface. A piece of material ¾" to 1" thick is usually adequate, and it should be clamped so it's perfectly level with the edge of the drawer front. Adjust the bit depth so it meets the scribed line on the inside of the drawer front as shown on page 113, turn on the router, and nibble out the waste between pins.

If you're careful, the bit can be brought surprisingly close to the scribed pin lines—less than ¹⁄₁₆". Work slowly, be mindful of the direction the bit is spinning, and keep an eye on the back line as well as the pin line. The bit will leave rounded corners, but it will remove almost all of the waste.

CLEAN UP THE CUTS AND ASSEMBLE THE JOINT

Once the router has finished its work, what you should be looking at is a nearly complete joint. All that remains is to pare away the little bit of remaining material on the drawer front pins with a sharp chisel.

Rather than removing all the waste in one swipe, it's best to take smaller bites. Simply place the chisel just a bit back from the routed edge and tap it lightly to pare away a sliver of wood. Work up to the scribed line, finally placing the edge of the chisel right in the scribe line (this is why it's better to use a layout knife rather than a pencil). A source of raking light will make the lines easier to see,

Now transfer the tail layout to the edge of the drawer front by aligning the drawer side on the scribed line on the drawer front and marking the pins with a knife.

Holding the drawer side firmly, use a layout knife to scribe lines on the edge of the drawer front. These lines mark the dovetail pins.

POWER-CUT DOVETAILS | Power-Assisted Half-Blind Dovetails

Get ready. Clamp the drawer front at a comfortable working height, with the inside of the drawer facing you. To help stabilize the laminate trimmer, clamp a scrap of wood to the drawer front. Make sure the clamps are set below the base line.

Rout away. Adjust the base of the router so the end of the bit just touches the base line. This carbide spiral bit removes waste cleanly. Turn on the router and nibble out the waste between pin lines. Work slowly and be careful not to drift over your layout lines.

but you can also darken them in with a pencil. You'll still be able to feel the edge of the chisel slip into position when you make the last cut.

Use the same approach to pare away the waste on the back line, and work carefully into the corner, taking a little bit of material off at a time until the corner is completely cleaned out. One thing to watch for is wild grain—it can send your chisel off in the wrong direction and damage the workpiece. If working straight down seems to be causing a problem, try paring in cross grain from the side. If you do shear off a section of pin (and it happens occasionally), you can always use cyanoacrylate glue and an accelerator to quickly repair the damage.

With both pins and tails cut, the joint can be tested for fit. Ideally, it will take light taps of a mallet or small hammer to bring the two pieces together. Getting a good joint is a matter of practice and feel. If the saw cuts are square and the marks transferred accurately to the drawer front, joints made this way should fit together nicely. Start the joint together to check—you can always make a minor adjustment now if something seems too tight—but don't put the joint all the way together just yet. Once you're reasonably sure the joint is going to fit together to your liking, add glue to the sides of the pins and assemble your parts. When the glue is dry, trim the joint flush. The result should look hand-cut because it is hand-cut. You've just had an assist from a power tool perfectly suited for the job. ∎

Power-Assisted Half-Blind Dovetails | **POWER-CUT DOVETAILS**

After routing. When you're finished with the router, there should be very little material left to remove.

Clean up. Begin paring away the waste on the sides of the pins. Take small bites. Finish each pin by placing the edge of the chisel right in the knife line and tapping lightly with a mallet.

Clean the back. When the sides of the pins have been pared to the line, work on the front of the joint. Make sure the front edge and the sides of the pins are square, not sloped, so the joint will go together without splitting the drawer front.

Assemble. First test the joint by just starting the two pieces together. Pare away any material that would seem to interfere with the joint. Then apply a bit of glue to the sides of the pins and assemble the joint.

You're done. Parts should fit snugly, but if you find you have a gap between a tail and a pin, a sliver of wood and a touch of glue will work wonders.

ROUTER TABLE DOVETAILS

Two sleds combine the ease of routing with the flexible spacing of hand-cut dovetails.

BY BRAD HOLDEN

There are lots of excuses for not using dovetails; cutting them by hand takes time, patience, and experience. A dovetail jig is relatively foolproof, but if you don't use it regularly, be prepared for an hour or so of re-learning each time you use it. Also, dovetail jigs are expensive, particularly those that are capable of cutting dovetails with variable spacing to give a hand-cut look.

This router table method lays waste to all the excuses. It's easy, accurate, and inexpensive. All you need are two bits and two sleds for your router table. Layout and cutting is straightforward, and you can space the dovetails any way you want!

MAKE TWO SLEDS

The two sleds (p. 123 and p. 124) must be custom-fit to your router table, so use the Cutting List on page 126 as more of a parts index. As long as your router table has two parallel sides, you're good to go. Cut the bases the same size as your router table's top, adding 1½" to the width. The extra width allows you to attach the rails to the underside of each base. When you

Router Table Dovetails | **POWER-CUT DOVETAILS**

attach the rails, squeeze them against the sides of your router table slightly (bottom left). They should be snug enough to eliminate any play, but still allow the sled to move easily.

Use your router table to cut a slot in each sled (bottom right). For the tail sled (that's the one you'll use with the dovetail bit), make the slot wide enough for the flared end of your dovetail bit to pass through. For the pin sled (that's the one you'll use with a straight bit), cut the slot using the same bit you'll use for cutting the pins. This slot is really the key to the accuracy of this method because it shows exactly where the bit will cut.

MAKE THE FENCES

These fences are just giant wooden T-tracks. You'll need three of them. Use a dado set or router to cut the wide groove in the fence body (p. 125) and then glue on the keeper rails. When the glue is dry, check to make sure the fences are flat and square. If they're not, take the time to true them up on your jointer.

Next, mill one long board for the sliding clamp blocks and cut both blocks from it. Attach a toggle clamp to each block. You only need two clamp blocks. You'll move them from fence to fence as you work.

Loosely clamp one fence to the tail sled. Use a square to align the fence exactly perpendicular to the sled's slot, then tighten the clamps (p. 122, top left). Turn the sled upside down and screw the fence in place. Don't use glue, just in case you need to make adjustments.

The pin sled uses two fences, both set at the same angle as your dovetail bit. You can use any dovetail bit you

Sled to success. With these two sleds, cutting dovetails with the router is a snap.

Side rails. Attach side rails to the bases of the pin and tail sleds. These rails allow the sleds to slide without any side play.

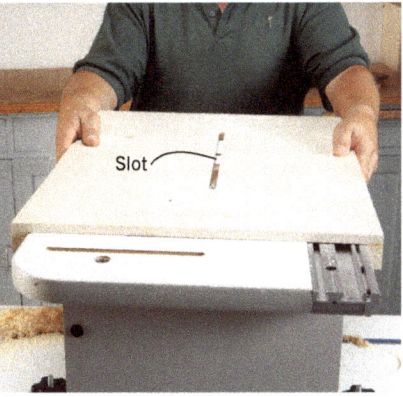

Make a slot. Rout the pin sled's slot with the straight bit you'll use for routing the pins. Rout the tail sled's slot wide enough to allow clearance for the dovetail bit's flared end.

POWER-CUT DOVETAILS | Router Table Dovetails

Square the fence. Square the tail sled's fence to the slot. Clamp it in place and then attach it from underneath using screws.

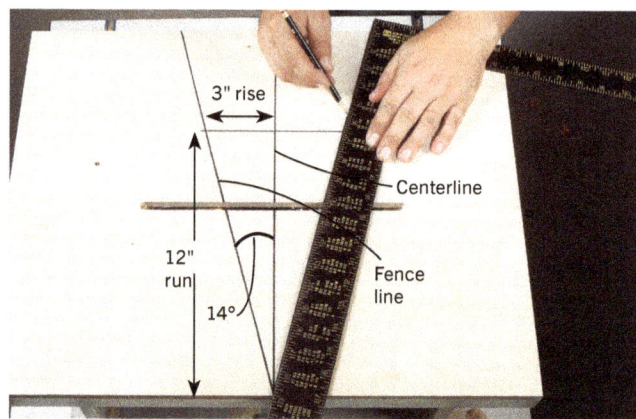

Angle. Lay out the angles of the pin sled's fences. Use a rise-over-run method to match the angle of your bit. In this case, the angle is 14°, which neatly works out to 3" of rise over 12" of run.

Get ready. Clamp the fences on the layout lines and attach them with screws. With both sleds complete, you can begin routing. Start by making the tails.

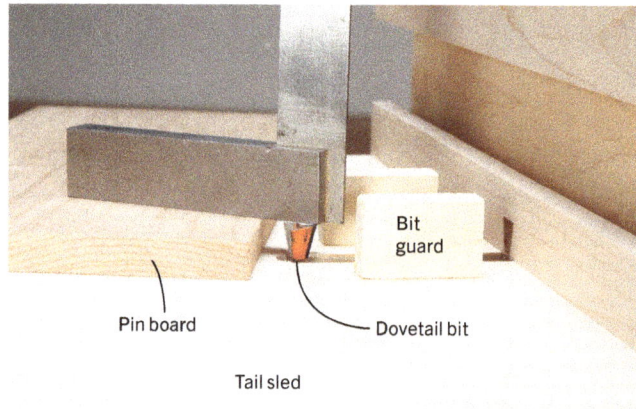

Tail settings. On the tail sled, set the dovetail bit's height to match the thickness of the pin board.

like. A shallow angle—anywhere from 6° to 8°—looks the most like hand-cut dovetails. I used a 14° bit because that's what I had on hand.

I used the rise-over-run method to mark the angles on my sled, as it's more accurate than a small protractor. If you don't know how to convert an angle to rise over run, search the internet for a converter.

To use your rise-over-run figures, mark a centerline across the sled, 90° to the slot. Make a mark on the centerline 12" from the sled's edge.

Make another mark 3" out from this point, 90° from the center line. Connect that mark with the end of the centerline for a perfect 14° angle (top right).

Cut the fences to length and clamp them to the base, one on each angled line (center left). Attach the fences with screws, as you did the tail sled's fence. Glue the bit guards to both sleds in front of each fence, on both sides of the slots. Leave enough clearance between the fences and the bit guards for your stock. The bit

Router Table Dovetails | **POWER-CUT DOVETAILS**

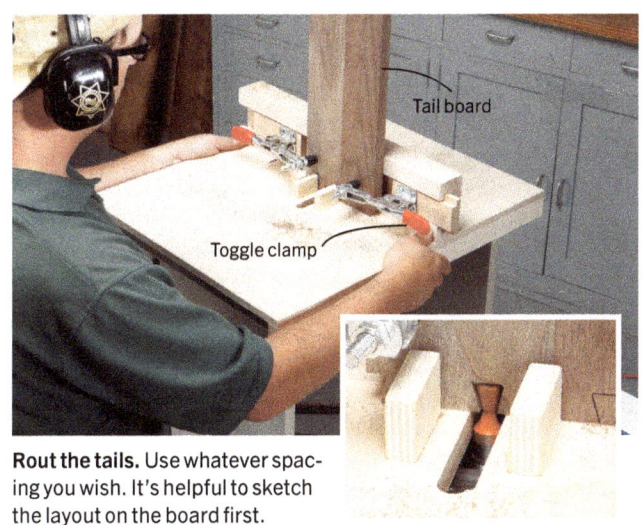

Rout the tails. Use whatever spacing you wish. It's helpful to sketch the layout on the board first.

Transfer the marks. Mark the tails on the pin board's end. These marks are just a handy visual reference so you don't get confused when you're routing the pins.

guards add a small degree of safety, but they're mainly a reminder to keep your fingers out of the danger zone.

CUT THE TAILS

Install your dovetail bit (I recommend using one with a ½" shank, to minimize vibration) and place the tail sled on your router table. In order to raise the bit sufficiently above the sled, you probably won't be able to bottom out the bit in your router's collet.

Set the bit's height to your stock's thickness using a straightedge (p. 122, center right). If your pin board and tail board are different thicknesses, set the bit's height to the thickness of the pin board. That's it; you're ready to start routing tails.

The beauty of this method is that you can space the tails however you wish. You don't have to mark the tails on the board, but I like to sketch them in so I can see what the finished joint will look like.

I cut the outer half-pin spaces first, then the interior-pin spaces, but the order isn't important. Position your tail board, clamp it to the fence, and rout each pin space to create the tails (top left).

CUT THE PINS

Clamp the tail board in a vise with its end extending above the bench's top

EXPLODED VIEW: TAIL SLED

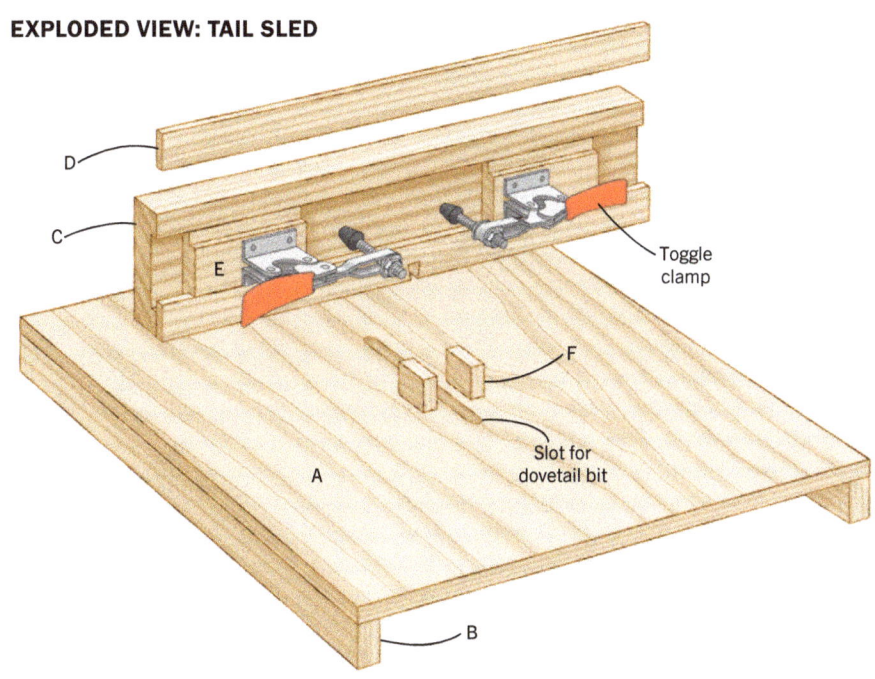

THE DOVETAIL BOOK | 123

POWER-CUT DOVETAILS | Router Table Dovetails

Continue marking. Extend lines from the edges of each tail onto the pin board's outer face. These are your layout marks for routing.

End flush with face

SQUARE IT UP

Traditionally, dovetails were cut slightly proud and planed flush after assembly. This isn't necessary here. Just clamp the pin board's face flush with the tail board's end. Now you can use a square to transfer your marks directly from the tails (left).

3D VIEW: PIN SLED

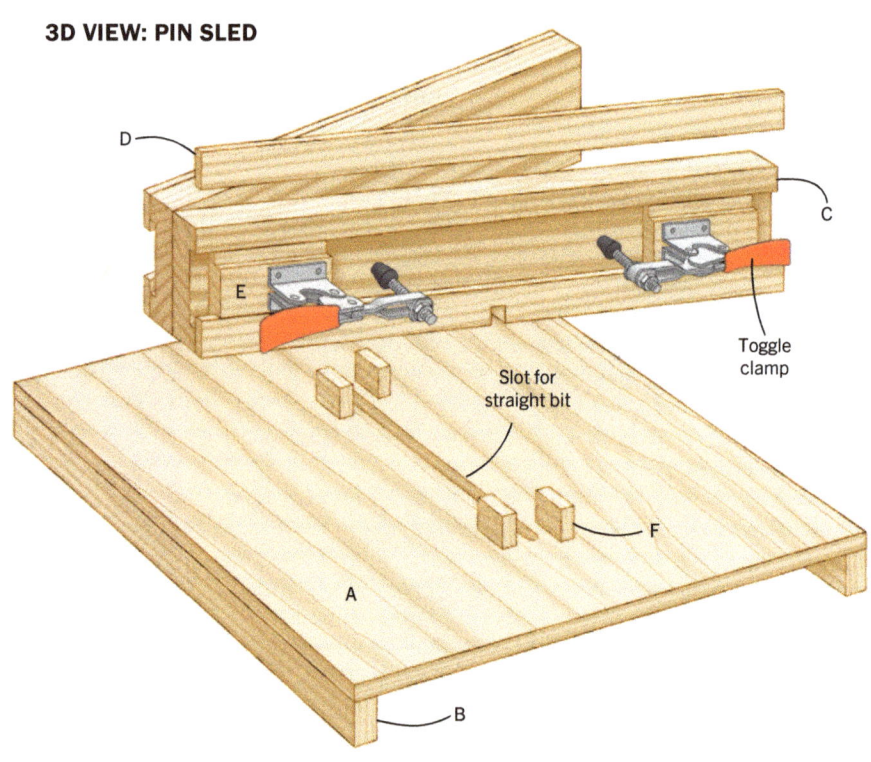

Toggle clamp

Slot for straight bit

at the same height as a spacer block. Slide the spacer block to the pin board's end, clamp the pin board near its center, and mark the dovetails on the pin board's end (p. 123, top right). Strictly speaking, these marks aren't necessary, but they're nice to have as a reference when you're orienting the pin board on the sled. Make a mark on each board to indicate its outer face.

With the pin board and tail board clamped in place, transfer the edges of the dovetails to the pin board's outer face.

Install the bit you used to cut the pin sled's slot, and swap the tail sled for the pin sled. Set the bit's height to the tail board's thickness.

Since you only mark the pin board's outer side, that's the side you'll have facing out. Draw symbols on the sled indicating the correct orientation of the bit and pin. This way, you'll always know on which side of the slot to position the pins. If the pin is to the left of the slot on one fence, it'll be to the right of the slot on the other fence.

Starting on either fence, align each pin's edge mark with the edge of the slot and rout one pass. To start

Router Table Dovetails | **POWER-CUT DOVETAILS**

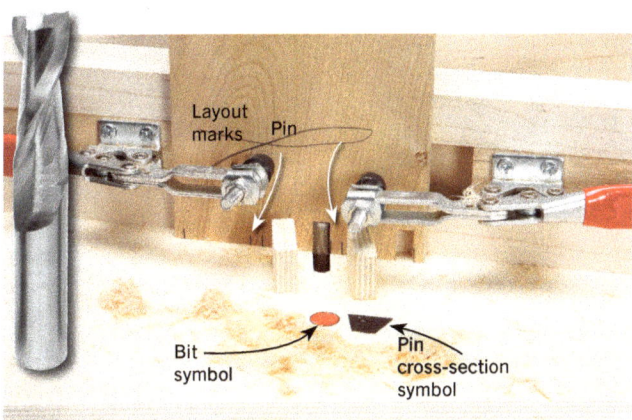

Pin sled. On the pin sled, align the layout marks with the slot's edge and rout one side of each pin. Mark the pin/bit relationship in front of each fence to ensure that you rout on the correct side of the pins.

Check your cuts. Check against the layout lines on the pin board's end to make sure you've cut in the correct places—and in the right direction!

with, cut a little wide of the mark and sneak up on the line. Check the cut against your tail board after each cut. After you rout a couple pins, eyeballing their alignment with the tails, you'll quickly learn whether to leave your layout marks visible, cut them off, or split them down the middle. A perfect fit is the goal, so leaving the pins slightly large for final adjustment is OK.

After routing the first side of all the pins, check to make sure everything is oriented correctly (top right). Now move to the opposite fence and rout each pin's other side. Rout out

DETAIL: FENCE

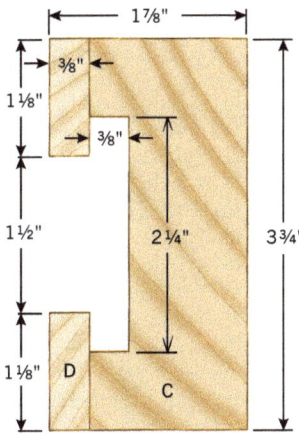

DETAIL: SLIDING CLAMP BLOCK

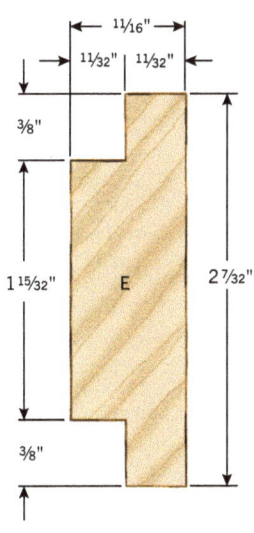

THE BEST BIT

A spiral up-cut bit is perfect for use on a router table (top left). An up-cut bit pulls the workpiece toward a router's base; in this case, that's down toward the sled. This makes the workpiece less likely to shift around, so your cuts will be more accurate.

POWER-CUT DOVETAILS | Router Table Dovetails

Reverse. Rout the other side of each pin using the sled's opposite fence. Here, the pin/bit symbols are reversed. After making one cut next to the pin, rout out the rest of the waste.

Tap the joint together. This shouldn't require much force. If the fit is tight, mark the pins that are too wide and trim as needed, using the pin sled.

any remaining waste by sliding the pin board over and routing another pass, repeating until all the waste is cleaned out (top left).

ASSEMBLE THE JOINT

The joint should fit with a few light mallet taps (top right). If you were cautious, cutting the pins just slightly oversize, you'll have to trim a couple of them. When you test the fit, mark the pins that need trimming. Because of the zero-clearance effect of the bit's slot, you can easily shave off just a whisker for a perfect fit. As you gain experience, you'll spend less time trimming to fit. ■

CUT LIST & MATERIALS

	QTY		ITEM	DIMENSIONS (INCHES)			MATERIAL
				T	W	L	
☐	2	A	Base	½	17 ½	22*	Birch
☐	4	B	Rail	¾	1 ¼	22"	Melamine
☐	3	C	Fence body	1 ½	3 ¾	20†	Maple
☐	6	D	Fence keeper rail	⅜	1 ⅛	20†	Maple
☐	2	E	Sliding clamp block	11/16	2 7/32	4	Oak
☐	6	F	Bit guard	½	1	1 ⅝	Birch plywood
☐	2		Toggle locking clamp				Such as DESTACO

Note: * Cut to fit your router table. † Cut length to fit sled.

SLIDING DOVETAILS WITH A ROUTER

Two router bits with guides and a simple shop-made jig make three variations of this joint a snap.

BY GLEN HUEY

One of the defining features of 17th- and 18th-century furniture is the dovetailed horizontal case divider. Case dividers are the rails that separate the drawers, or the door and drawer sections. Attaching these dividers to a case's sides using sliding dovetails is probably the strongest way possible to assemble a carcase.

POWER-CUT DOVETAILS | Sliding Dovetails with a Router

However, reproducing this detail is daunting to many woodworkers. Not only is a sliding dovetail seen as complex joinery, but it can be made in different ways. The basic sliding dovetail, shouldered sliding dovetail, and through-sliding dovetail (shouldered or not) are just a few of the options.

Each type of sliding dovetail requires a different jig. I've used a variety of these jigs in my many years of building reproduction furniture. Some jigs capture the router base and are specific to a certain router bit. If you need to use more than one bit (to make a shouldered dovetail, for example) this can be a problem—unless you own two identical routers.

Other jigs are as large as the entire case side, making them hard to handle and store. I've found a better way. Using a ¾" top-bearing flush-trimming bit (often used for pattern routing), a ¾" dovetail bit, a template guide with a ¾" outside diameter, and a shop-made straightedge, any of these joints can be made easily.

FROM DADO TO DOVETAIL

To understand how this works, let's start with a simplified version of the joint: a dado. With a straightedge clamped across a cabinet side and a flush-trimming bit in your router, you can cut a dado for case dividers or web frames. Simply position the straightedge where you want your dado, set the depth of cut on your router, and plow it out. The bearing on the bit follows your straightedge.

By using a dovetail bit with a template guide and this same setup, you can use the straightedge to make the basic sliding dovetail shown at top.

DETAIL: BASIC SLIDING DOVETAIL

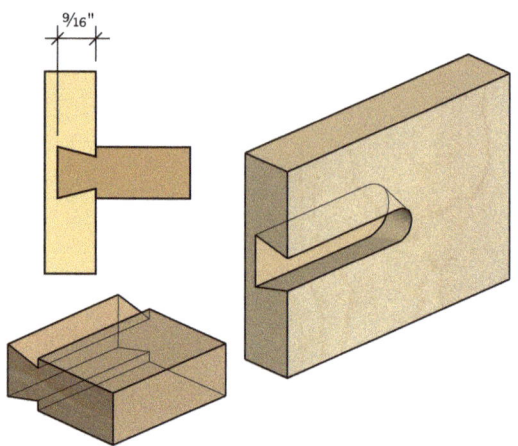

The simplest option in sliding dovetails. The socket or trench requires only a single pass with the dovetail bit.

DETAIL: SHOULDERED SLIDING DOVETAIL

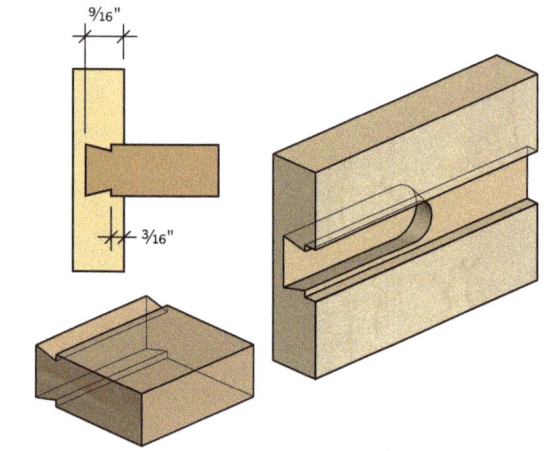

This joint adds a shoulder to the dovetail and requires you to make a first pass with a pattern-making straight bit. Then the dovetail bit (set at the same depth as on the joint above) is used to cut the dovetail.

DETAIL: THROUGH-SLIDING DOVETAIL

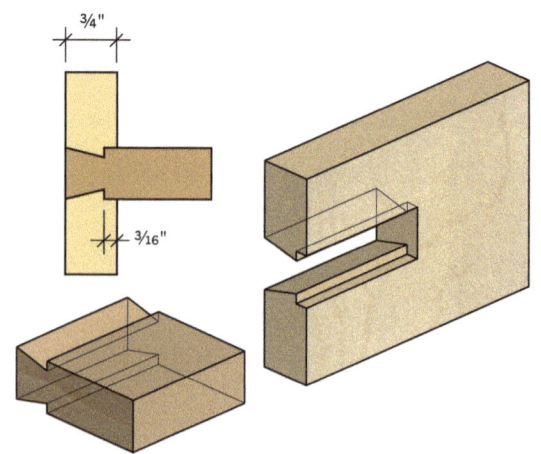

This more-complicated joint is made in three steps. First the straight bit forms the shoulder, then the dovetail bit shapes the divider pin. The final dovetail socket is hand-cut to avoid tear out.

Sliding Dovetails with a Router | **POWER-CUT DOVETAILS**

THE STEPS TO A SHOULDERED SLIDING DOVETAIL

Making a shouldered sliding dovetail begins by cutting a dado in the case's side. This dado is easily made with a pattern-cutting bit and the right jig, which I call a straightedge guide.

The bed of my jig, shown below, is simply two pieces of plywood cut slightly longer than the width of the case side, then glued or screwed together face to face. (Depending on your router and bit, you might need only one thickness.) To complete the jig, screw a third block to the underside of the straightedge guide to hook it square against the front edge of the case side. The hook should be sized so you can clamp the jig in place without interfering with the base of the router. As you cut the dado, make sure you move the router in the correct direction (against the rotation of the bit) to keep it tight against the jig.

Next, install a template guide in your handheld router and the dovetail bit. I should mention one important detail: To use a template guide that is the same diameter as the pattern-cutting bit's bearing collar (in this case ¾"), it will be necessary to attach the guide first, then insert the bit afterward. Because of the identical diameters, the router base can't be slipped over the bit with the template guide in place. The guide is the same diameter as the collar to allow the dovetail to run exactly down the center of the dado cut.

With the template guide in place and the depth set on the dovetail bit, you're ready to cut the dovetail socket, as shown below center.

With the socket created, it's time to make the mating tail on the end of the drawer divider as shown below right. Mount the dovetail bit in a router table and run both sides of your divider on end between the fence and bit. You will need to make a few test passes to get the perfect fit. Note that I'm using a push block behind the divider for safety and to stabilize the piece during the cut.

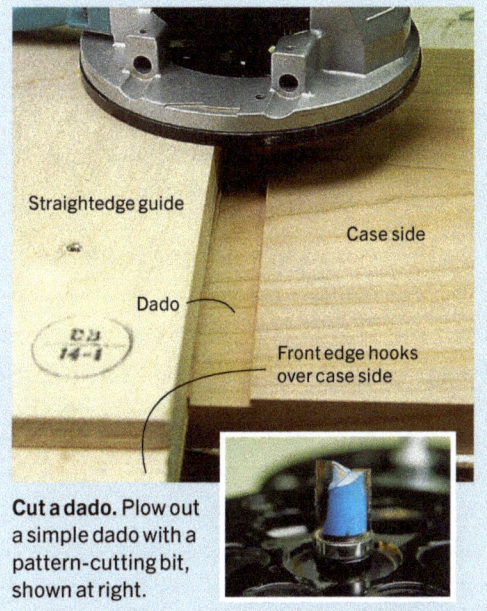

Cut a dado. Plow out a simple dado with a pattern-cutting bit, shown at right.

Cut the socket. Use a dovetail bit, shown at right, to make your shouldered dovetail socket.

Cut the tail. You can easily rout the tail of the joint on your router table with the matching dovetail bit.

POWER-CUT DOVETAILS | Sliding Dovetails with a Router

Use a template guide that has the same outside diameter as your dovetail bit to make measuring simple. Next, clamp your straightedge exactly where you want the sliding dovetail to go.

Set the proper depth for the bit, (9/16" in 3/4" material, for example) then rout the dovetail trench or socket in a little further than the width of the divider. The trench doesn't need to extend all the way across the side. Because the dovetail trench will have a rounded end, the trench must extend a little further so the square-shouldered tail on the divider will fit.

TWO-STEP SHOULDERED JOINTS

A shouldered dovetail is ideal for casework that uses web frames to support drawers. The straight shoulder, which supports the web frame, is cut just as you would cut a basic dado.

First align your straightedge as you did with the basic sliding dovetail. With a 3/4"-diameter flush-trimming bit in your router, plow out the dado to 3/16" deep. Next, take your router with a template guide and dovetail bit, set it to 9/16" deep (without moving the straightedge), and make the cut into the case side. The cut should be a bit longer than the width of your front divider.

Thanks to the template guide (and keeping the straightedge in one fixed location), the dovetail portion of this cut is centered in the dado automatically.

THROUGH-SLIDING DOVETAILS

For an even fancier look, you can create through sliding dovetails. These joints allow the end of the case's divider to be seen on the outside of the case.

Start once again by plowing the dado as explained earlier. You could cut the socket portion of this joint with a router, but there's much less chance of tear out if you cut the socket using a handsaw.

If you go with this hand-tool route, you should first cut the male portion of the joint (called the tail) on the end of your horizontal divider using the dovetail bit in your router table. The process is explained below. Then use the tail to lay out the location of the socket on the case side.

Now you can saw out the socket. Orient the saw to match the two tail sides, then cut in from the front edge the width of the divider. Finally, chisel out the waste between your saw cuts.

DON'T FORGET THE TAILS!

To make the mating joinery on the dividers (the tails), I use my router table. Use the same dovetail bit you used to cut the dovetail sockets to form the tails to ensure that the joint fits well. Set the fence to adjust the size of the tails, cutting on both sides of the divider. I like to sneak up on the final cut to ensure a snug fit.

Set the bit to cut at the appropriate height for each joint style. For the basic sliding dovetail, that height should be about two-thirds of the width of the case side. If you're making a shouldered dovetail, allow for the 3/16" shoulder depth in your layout.

The through-dovetail is cut with the height of the tail equal to the thickness of the case side (if you are adding a shoulder, remember to allow for the shoulder).

Your through-dovetail doesn't need to expose the whole width of the divider. For example, you can show only 3/4" on the sides if you like. After cutting the tails on both ends of the divider, use a saw to trim the end 3/4" back from the front of the divider on both sides. Then cut from the back of the divider right at the point where the tail begins from the divider to remove the unneeded tail section. Repeat this cut on both ends.

With the back portion of the tail removed, slide the divider into the dado in the case and mark, then cut, the matching socket.

WHATEVER SIZE YOU NEED

While these techniques work well with the standard 3/4"-thick drawer dividers that are common today, they also work with other thicknesses of dividers by using different-sized template guides and bits. The guides are readily available in a wide variety of sizes, including 51/64" and 1", if you need thicker drawer dividers.

You should consider using sliding dovetails for any number of woodworking tasks. The possibilities are endless. ■

MAKE A SLIDING DOVETAIL AT THE TABLE

These tips and tricks at the table saw and router table help to get the perfect fit.

BY CHAD STANTON

Making sliding dovetail joints is craftsmanship at its best. You need decent tools, an eye for precision, and lots of practice.

In this article, you'll learn a sure-fire method for getting fantastic results. I'll describe the general principles involved and show you how to set up your machines. I'll also point out where you might get into trouble—and how to avoid it. If you haven't made a sliding dovetail before, this sample joint is a perfect place to start.

POWER-CUT DOVETAILS | Make a Sliding Dovetail at the Table

CROSS SECTION: JOINT

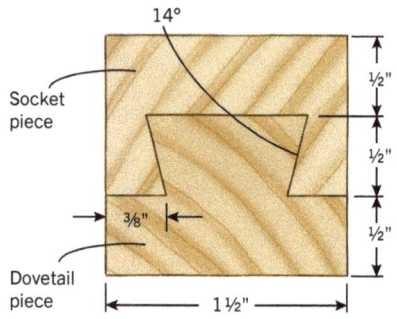

DETAIL: SAWING AND ROUTING STEPS

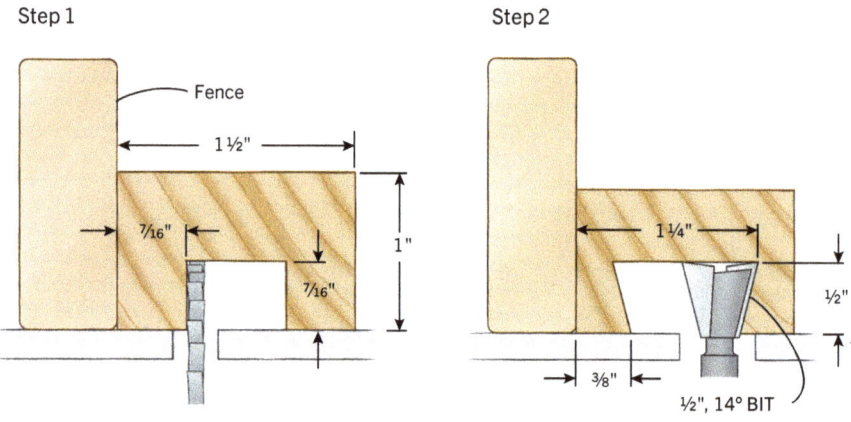

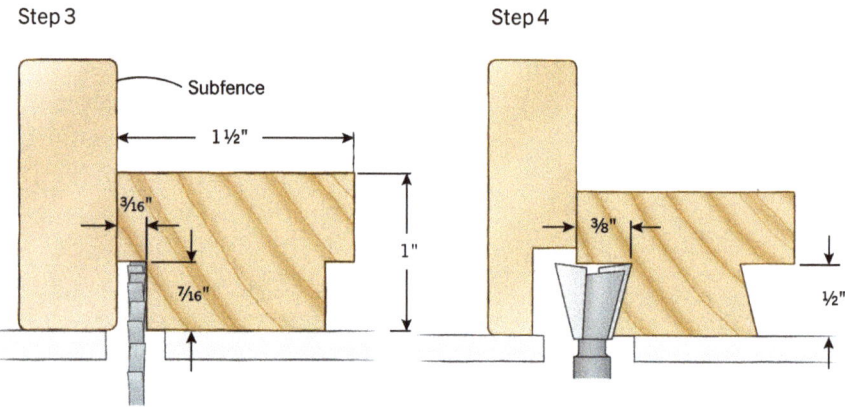

SET THE TABLE

For this exercise, start by milling a clear piece of wood 1" x 1½" x 36". You'll be routing this piece from end to end in both directions; to avoid tear out, the wood's grain should be straight and parallel to the piece's edges.

Your stock must be absolutely straight and flat, so it's best to mill it about ⅛" to ¼" oversize in thickness and width first. Set the piece aside a few days to bend or twist, if it wants, then flatten it again and mill it to final dimension. Crosscut the piece into three 12"-long sections.

One of these shorter pieces will be the socket half of the joint—you'll make this one first. The other two pieces will be used to make the "dovetail" half of the joint. Of course, you really only need one dovetail piece, so why make two? The second is for test cuts. Trust me, you'll need it.

Whenever you make sliding dovetails, it's a good idea to draw the joint full size first in order to work out its proportions (see top). Proportions are primarily a matter of taste, but here are a few practical considerations:

- *Watch the width.* Don't make the narrowest sections too thin and weak. Determine their width, using your best judgment, based on the strength of the wood and the amount of stress the joint will face.

- *Know the angle of your bit.* Angles range from 7° to 14°. (I used a ½" 14° wide bit, the standard for half-blind dovetail jigs). The smaller the angle, the more precise the fit must be.

- *Have a wide base.* Make the base of the dovetail piece fairly wide. If the dovetail is too narrow, balancing the piece on the router table is difficult

Make a Sliding Dovetail at the Table | POWER-CUT DOVETAILS

(see p. 134, bottom). A wider dovetail offers more support.

MAKE THE SOCKET

Draw the socket on the end of one of your pieces. Usually, there's a lot of waste to remove in making a socket, so I prefer to start on the table saw. Sawing is much faster than routing and easy to set up. This method also results in less wear on the dovetail bit; the sharper the bit, the easier it is to get a good fit.

Adjust the saw's blade and fence to cut $1/32$" to $1/16$" inside the outline of the socket (top). This joint is centered on the workpiece, which makes all the cutting and routing operations more efficient. Each fence setting is a twofer: After making one cut, rotate the piece and make another cut. Move the fence a couple of times, making overlapping cuts, until the groove is complete.

Next, set up the router table with the dovetail bit (bottom). I'll bet you'll be tempted, as I was at first, to position the fence close to the bit—but you'll be rewarded with an unpleasant surprise. If you're not careful, the bit will jerk your workpiece forward. Why? Well, take a look at the bit's rotation: Set up this way, the side that's doing all the cutting is pulling the workpiece. (Imagine feeding a table saw from the back side—yikes!) The solution is simple. Adjust the fence so that the bit is cutting the opposite side of the socket; now the bit pushes against the workpiece as you feed it across the router table—the correct way.

Thanks to the table saw work, you're removing very little wood in this operation. Go right for the lines you drew. After making one cut, turn the piece around to complete the socket.

Start with the socket piece. Remove most of the waste on the table saw, using a standard blade or a dado set.

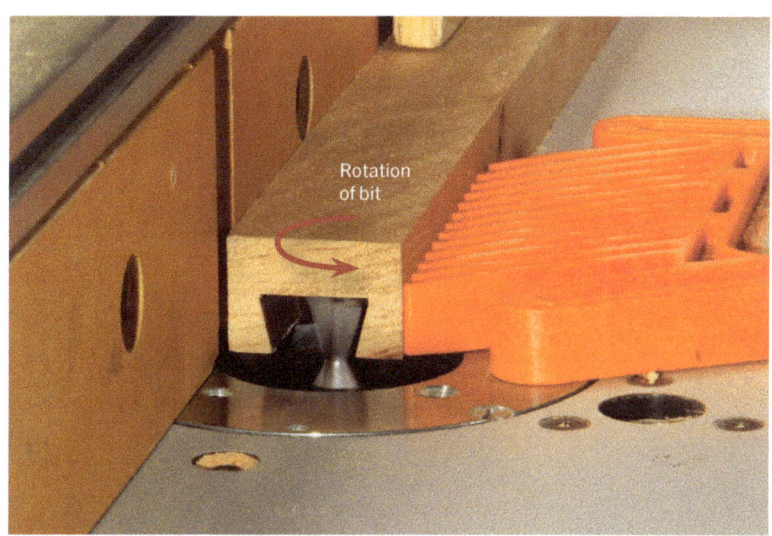

Rout the socket. To prevent the workpiece from being yanked forward by the bit's rotation, position the fence so the side of the bit cuts only the featherboard side of the socket, as shown. Flip the workpiece around to cut both sides.

MAKE THE DOVETAIL

Mark one of your dovetail pieces as "Test," then stand it on end in a vise. Place the socket piece on top of it and

POWER-CUT DOVETAILS | Make a Sliding Dovetail at the Table

Table saw to remove waste. Remove most of the waste from the mating piece. Use a subfence so you can position the blade right at the edge of the workpiece, creating a rabbet.

Cut the dovetail. Set up the router table for cutting the dovetail. Use the socket piece to position the fence. Support the piece on a board whose thickness equals the height of the bit.

trace the socket on the end of the test piece—this is just a rough guide. Back on the table saw, remove most of the waste on the test piece and on the "real" piece (top).

It's time to go back to the router table—but hang on a sec. Tear out above the bit is a common problem when routing the dovetail. The result is a raggedy edge on the side of the joint. If this surface will show on whatever you're making, it's best to score the wood first. There are two ways you can do this. One option is to use a cutting gauge (a marking gauge with a knife-like cutter) to incise a deep line where the dovetail's shoulder will be. Alternatively, begin routing by adjusting the router table's fence so the bit makes a cut that's only 1/16" or so deep.

After the scoring cut, adjust the router table's fence to make the dovetail full-depth. Don't go for it on one shot, however. The iron rule for this setup is to start fat and finish lean. That is, make the dovetail extra-wide at first, then reduce the dovetail's width by a series of light cuts until you get the fit you want. When you get close, a few thousandths of an inch matter—it's that fussy!

Without changing the height of the bit, use the socket piece to position the fence. Place this piece directly above the bit, then move the fence so that the bit's tip extends 1/32" to 1/16" short of the portion of the socket immediately above it (left). Rout both sides of the test piece, then try fitting it into the socket—if all is well, it shouldn't go. Rout both sides of the second dovetail piece.

To make the dovetail smaller, gradually adjust the fence back from

Make a Sliding Dovetail at the Table | POWER-CUT DOVETAILS

the bit. I use a micro-adjust system that's dirt simple (top). There's nothing to build—you just need a clamp, a block, ten playing cards, and a couple of small pieces of paper. Place the cards and paper behind the fence; they'll act as shims. Push the block against the stack of cards and paper, then clamp the block to the router table.

To set up the next test cut, remove one or two cards, then loosen the fence and push it against the remaining cards. Tighten the fence. Rout one side of the test piece and try fitting the piece into the socket; if the dovetail is still too large (and it should be), rout the other side and try again. If it doesn't fit, fine—you're good to rout the second, "real" piece.

Continue in this fashion until the fit is just about right; at this point, you should be removing only one card at a time. When you get extremely close, take even lighter cuts by removing the pieces of paper, one at a time. Just remember to rout the pieces in tandem; do the test piece first, try it out, then do the real piece. Be patient. When you get close, removing wood from just one side of the dovetail might do the trick.

When the parts fit, plane or joint the bottom of the dovetail piece to minimize friction between the two pieces (bottom).

Hang on to the sliding dovetail pieces once you're done. They'll be good to refer to later, when you need to make an actual joint. ∎

Rout the dovetail. Don't go for the perfect fit on the first pass; start oversize, then move the fence back in small increments to make the dovetail narrower. Place several shims between a stop block and the fence. After routing both sides, remove one shim and reposition the fence.

Shave a bit. Plane a few shavings off the bottom of the dovetail. This creates a clearance between the two halves of the joint, making it easier for them to slide past one another.

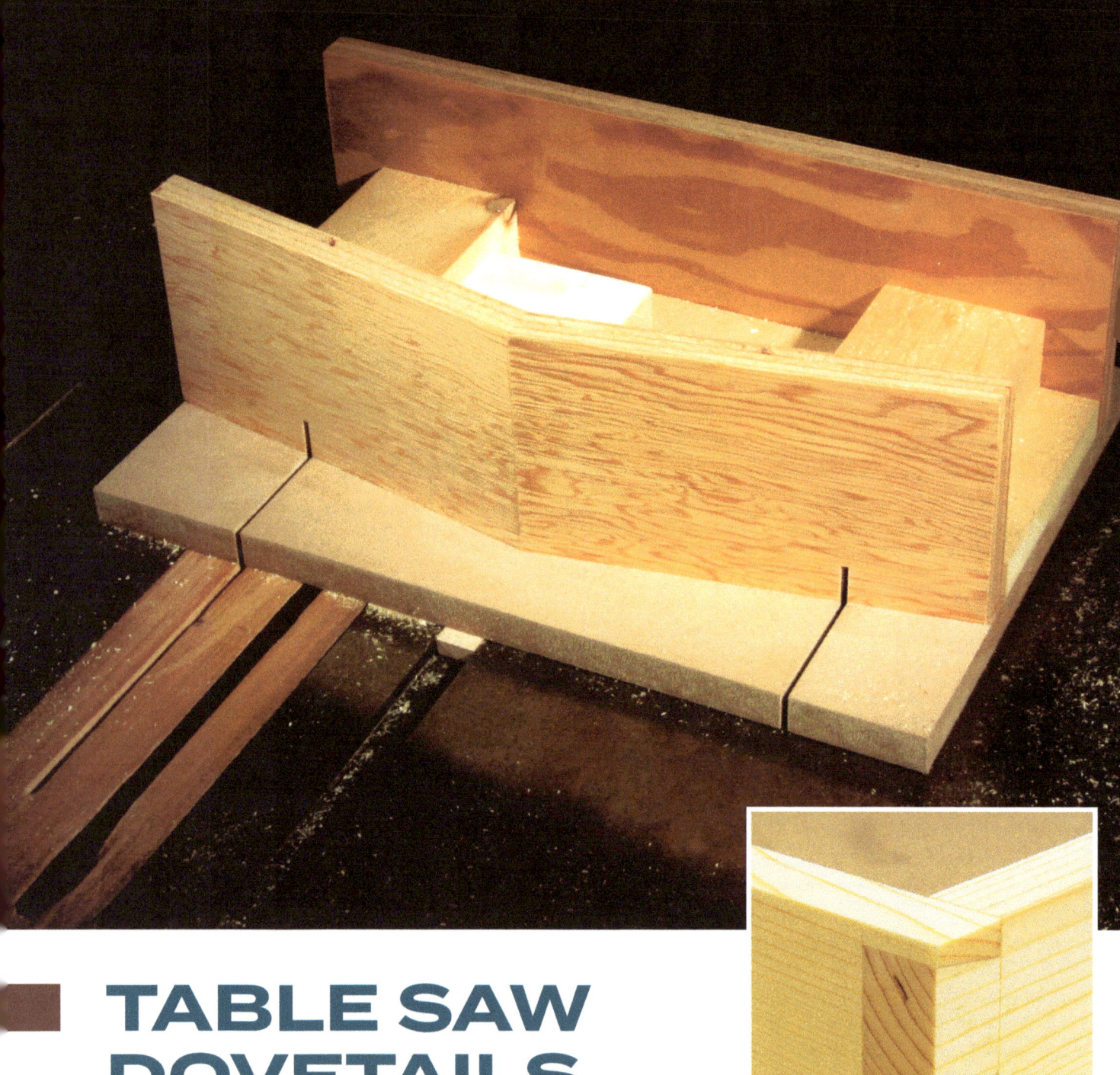

TABLE SAW DOVETAILS

These machine joints look handmade.

BY JIM STACK

Table Saw Dovetails | **POWER-CUT DOVETAILS**

Imagine cutting through-dovetails with pins and sockets spaced any way you like—as if they were cut by hand. Now imagine cutting them without using a dovetail saw or a router. The humble fixture showcased here allows you to do just that—using your table saw. It allows cutting both pins and tails that require only minimal cleanup, and it can be used to make unique pieces or production runs (simply clamp a stop on the fences to locate the identical parts). This fixture is easy to build and it can be adapted for use on virtually any saw. The first time I used this method to cut dovetails, they came out perfectly—and they have ever since. Am I that good? Nah, it's because this fixture is that easy to use.

BUILD THE FIXTURE

The fixture's length depends on two details: the distance between your saw's two miter slots and the width of the pieces it's designed to cut. This fixture is sized for use with slots spaced 10"–12" and workpieces up to 5" wide (see Exploded View and Cut List at right). Use a hardwood such as maple to make the guide rail; use pine, plywood, or MDF for the other parts.

Cut the base to final width and length. Make the cleat from a rectangular blank. On the leading edge, precisely cut 10° angles that meet at the middle. This angle determines the slope of the pins and sockets. (To change the slope, use a different angle.) Glue the cleat to the base, making sure its straight back edge is parallel to the back edge of the base.

Miter one end of each angled fence at 10°. Then cut both fences to length—it's OK if they're longer

EXPLODED VIEW: DOVETAILING FIXTURE

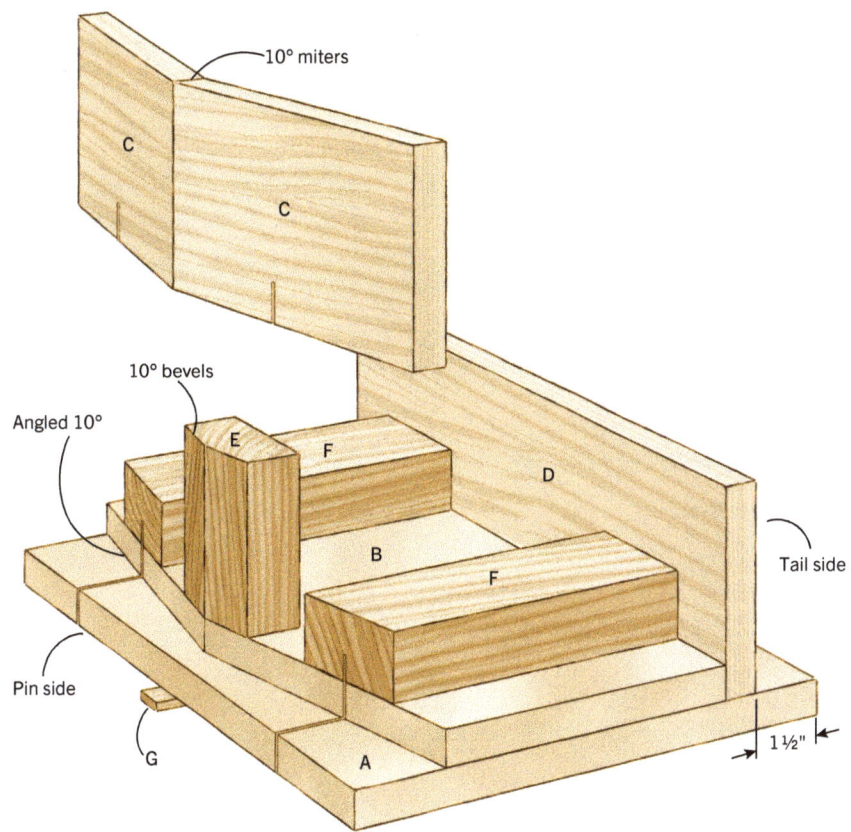

	QTY		ITEM	DIMENSIONS (INCHES)			MATERIAL
				T	W	L	
☐	1	A	Base	¾	11	16	MDF
☐	1	B	Cleat	¾	6 ½	16*	MDF
☐	2	C	Angled fence	¾	4 ½	9†‡	Plywood
☐	1	D	Straight fence	¾	4 ½	16	Plywood
☐	1	E	Glue block	1 ½	3	3 ¾§	Pine
☐	2	F	Blade guard block	1 ½	3	7‡	Pine
☐	1	G	Guide rail	¼	¾	12	Hardwood

Note: * Angle the front edge at 10° from both ends to the center. † Bevel one end at 10°. ‡ Cut length to fit. § Bevel the front face to fit behind the angled fence.

POWER-CUT DOVETAILS | Table Saw Dovetails

Locate the guide rail. The fixture must allow positioning of the blade on both sides of the angled fence. Find the sweet spot by marking a possible location for the rail and moving the fixture from one miter slot to the other.

Lay out the pins. Proceed as if cutting dovetails by hand. Their slope must match the angle of the fixture's fences. Mark the waste on the outside face (where the pins are wide).

Cut the first half pin. Use the scribed layout line to set the blade height. The fixture's angled fence automatically sets the slope. Clear the waste by making successive adjacent cuts.

than the base, but they shouldn't be shorter. Glue the fences to the cleat's angled edge, making sure they're square to the base—install shims, if necessary. Similarly, cut and glue on the straight fence. Shape the glue block to fit behind the angled fences. Clamp the block in your vise and have at it with a handplane. Then glue it in place. Cut, fit, and glue on the blade guards.

INSTALL THE GUIDE RAIL

Fit the guide rail to your saw's miter slots. The rail should fit snugly yet slide easily, without binding or wobbling. Locate the guide rail by roughly centering the fixture (with its angled fences facing the blade) on one of the table saw's miter slots (top). Mark the slot's location on the fixture. This mark represents a possible location for the guide rail. Move the fixture so the mark is centered on the saw's other miter slot. At both locations, the goal is for the blade to be roughly centered on either the right or left angled fence. Miter slot locations and their relation to the blade vary from saw to saw, so you'll probably have to try a few different guide rail locations to find the one that works the best. Mark this "sweet spot" on the front edge of the fixture's base.

Flip over the fixture and clamp one end of the guide rail to the base at your mark. Drill a pilot hole and install a ¾" #6 screw. With the fixture's straight fence facing the blade, slide the guide rail into one of the saw's miter slots. Then use the straight fence and a framing square to square the fixture with the saw's rip fence (which, of course, is paral-

Table Saw Dovetails | **POWER-CUT DOVETAILS**

lel to the blade). Hold the fixture in position and precisely mark where the rail extends beyond the back edge. Remove the fixture, flip it over, and use this new mark to locate the guide rail. Clamp the rail in position and install another screw. Test the guide rail's fit in both miter slots. If the fixture doesn't slide smoothly, rub paraffin wax on the rail and in the grooves. If the fixture still resists, use a rabbet plane to remove a tiny amount of wood from one side of the rail. Try the fixture again and repeat until the fixture glides smoothly in both slots. Remember that any side-to-side movement in the slots is bad.

CUT PINS

Start by laying out the pin board (p. 138, center). Use a scoring tool to mark the length of the pins—their length is determined by the thickness of the tail board. The scored marks will prevent tear out when you remove the waste between the pins. Use a sliding bevel square to mark the angled pins. Set the bevel to match the angle of the fixture's front fence (80° in this case). Mark the waste between the pins.

Place the workpiece on the fixture with its inside face against the angled fence—make sure it's on the correct side of the fence. Use the scored mark to set the blade height. I install a rip blade to make these cuts because its flat-topped teeth cut flat shoulders, virtually eliminating clean-up.

Make a cut on the waste side of the half pin (p. 138, bottom). Reposition the workpiece and make additional passes to clear the waste, stopping just shy of the line indicating the next pin. Repeat this process to cut

Continue. Jump to the same side of the next pin and repeat the process. Use a rip blade to make these cuts—its square-topped teeth leave flat shoulders.

Finish cutting the pins. Move the workpiece to the other side of the angled fence. This side slopes the cut in the opposite direction.

Time for tails. Use the pins to locate the tails. Use a sharp pencil, so there's no gap between the pencil mark and the pin.

THE DOVETAIL BOOK | 139

POWER-CUT DOVETAILS | Table Saw Dovetails

Cut the tails. Tilt the blade, turn the fixture around, and use its straight fence to cut the tails. Make all the cuts that angle one direction, then flip the workpiece to its opposite face to complete the job.

Remove waste. Square the shoulders at the scribe line, using a chisel.

the remaining pins that slope in the same direction (p. 139, top). Then move the fixture to the other miter slot to cut the pins that slope in the opposite direction (p. 139, center). Nibble out the waste as needed.

CUT TAILS

Use the pin board to mark the tails on the tail board (p. 139, bottom). Tilt the saw's blade to match the tails' slope (10°). Whether the blade tilts to the left or right doesn't matter. I switch from the rip blade to a general purpose blade to make these cuts. Both blades leave ridges at the top of the cut that have to be removed later, but the ridges left by the general purpose blade are smaller.

Install the fixture with its straight fence facing the blade. Place the workpiece on the fixture and raise the blade to meet the scored line. Next, carefully position the workpiece so the blade will cut away all but a sliver of the sloped line. The placement is critical because these angled cuts determine the joint's fit: Remove too little and the joint won't go together; remove too much and the joint will be too loose.

Make angled cuts on one side of each tail (top). Then flip over the stock to cut the other sides. Use a chisel to remove the waste that remains between the tails (bottom). Keep the chisel handy for fine-tuning when you test fit the joint. ■

HOW TO MAKE CONDOR TAILS

This ingenious method combines routers, bandsaw, and hand tools for big dovetails.

BY JAMEEL ABRAHAM

I know what you're thinking: "Another opinion on how to cut dovetails." I hear you. But this one's different. I promise. No back and forth over pins or tails first. No Rob Cosman vs. Frank Klausz. Well, actually, a little Klausz.

When I built my first serious workbench, I practically memorized Scott Landis' *The Workbench Book*, and like many woodworkers, I was attracted to Frank Klausz's beautiful bench, especially the large, crisp dovetails that joined the parts of the tail vise. Klausz told us what tools he used to cut the joints, but didn't elaborate much on technique. I suppose with a lifetime of skill at your command, you just pick up the tools and the joint emerges. I wanted the crisp look of Klausz's large-scale joints without waiting 20 years to develop the skill. After building several large benches over the past few years, this technique emerged.

BEST OF BOTH WORLDS
I'm a big believer in making dovetail joints that fit right off the saw. That's a skill that's easy to learn with some practice. But not so with the beefy members of a workbench, or large-

POWER-CUT DOVETAILS | How to Make Condor Tails

Leave a space. Size your tails so the base is wider than the router bit's diameter.

scale furniture. When you need to cut tails on the ends of an 8' board, how do you hold the workpiece? I've seen people stand on top of their bench, climb ladders, even clamp the board to a second-story deck railing just to get the thing vertical so they can use a backsaw to cut the joint. Instead of that drudgery, I lay the piece flat and use the bandsaw and router to cut this joint, utilizing the strengths of those machines. I also use hand tools where they excel. This is truly blended woodworking.

This technique uses the same sequence of layout and cutting as if you were making the joint by hand, and all the critical fitting is done by hand, using accurately scribed lines. The machines provide some precision, but none of the fit is dependent on super-precise machine setups. This technique works equally well with half-lap or through-dovetails.

LAYOUT

To get started, lay out the tails in the typical fashion. I'm cutting a half-lap (also called half-blind) joint for a workbench's front laminate (the tail board) where it joins the end cap. Both boards are 4" wide, with the tail board 1½" thick. Because this is a half-lap dovetail I set the marking gauge to leave about ⅝" of material on the pin board past the tails. This isn't a critical dimension, so I go for looks—beefy for a workbench. Scribe all the way around the board.

After scribing, use a bevel gauge and pencil to lay out the two tails. Set the gauge to about 7°.

Here's an important point: When laying out the tails, make the width of the tail's base about 3/16" wider than your router bit. Later, you'll be routing away the tail sockets in the pin board with this bit, so it needs to easily fit between the pins. My pin's base is about ¾" wide.

CUTTING TAILS

The tails are cut on the bandsaw using a foolishly simple angled spacer that takes about three minutes to make. Cut a piece of wood (I used a plywood offcut) to about 16" long and 5" wide. The dimensions aren't critical. Now take your bevel gauge and draw a line along one long

Tandem slide. Slide both the spacer and workpiece along the fence to make the cut.

Mirror image. Flipping the board cuts the opposite tail perfectly without measuring.

Opposing angle. After changing the screw stop, cut the opposing angle, flip and repeat.

edge of the spacer. I use a long steel ruler to "extend" the blade on the bevel gauge. Cut to this line on the bandsaw. If you've cut straight and true, you don't even need to bother cleaning up the edge.

Next, drive a screw near the end of the tapered edge and let it protrude about 3/8" or so. This will be the stop for the tail board.

To set up the cut, keep the tail board pressed tight to the screw and the edge of the spacer, then approach the blade to set the fence. (Make sure the screw stop isn't in the path of the blade.) Adjust the fence so the blade is on the waste side of the line. You don't have to be too fussy here. Just like for a hand-cut dovetail, we're going to use the tails as a pattern for cutting the pins. You will want to set up a roller stand to support long workpieces. To make the cut, slide the spacer and workpiece along the fence. Stop just short of the baseline.

To cut the outside edge of the other tail, simply flip the board over and repeat. Don't worry about nailing your layout line, this will automatically size and center the tails on the board. And also don't fret if your angle is off a tad (you can see in the pictures that mine is). It doesn't matter one bit.

To cut the opposing angle on the tails, remove the screw stop, flip the spacer end for end, and replace the screw in the opposite end. Cut the remaining edges, flipping the board as before.

Next, I remove the spacer and use the bandsaw to nibble away some of the waste from between the tails.

Next, move to the bench and use a backsaw to remove the waste from

Chop by hand. Chiseling the end grain allows controlled precision.

Flat shoulder. Use a wide chisel to test for flatness.

the half-pin area. Here you want to maintain an absolutely crisp arris, so don't saw right to the scribe line. Stay away from it just a little.

To chisel the end grain precisely, I drop the edge of the chisel into the scribe line and tap firmly once. This pops out a small amount of material. Do this on both faces of the board (look close; you can see those areas I've removed). Then I flip up the workpiece onto its edge and chisel away the shoulder with a series of cuts, using a 1/2" chisel. I find that if I try to cut the entire shoulder with the workpiece on edge, using a chisel that's wider than the thickness of the board, I will almost always cut past the scribe line on the faces of the board as the chisel reaches the inside bottom of the shoulder. Cutting away a little ledge on the faces first allows

POWER-CUT DOVETAILS | How to Make Condor Tails

A 17 percent rebate. Cut away about ¼" to form the rabbet.

Mark tight. Mark the pencil lines tight to the sides of the tails.

Shift right, scribe left. Scribe lightly. You don't want to make a canyon here.

me to establish that crisp arris, thus I can stay away from it as I chisel the center portion of the shoulder, with the workpiece on edge. This also allows me to focus my attention on keeping one scribe line crisp as I chisel, instead of all three. Using a relatively narrow chisel also allows more control and precision. I don't chop aggressively; rather, I make several lighter taps to maintain control of the chisel.

To make sure I don't have a hump in the middle of the shoulder, I check it with the back of a wide chisel. I rock it back and forth. It should click down positively on each arris as you do this. Chisel out the waste between the pins the same way, getting 95 percent of the waste out of the way before you take your final pass with your chisel registered in the scribe line.

I use a router to cut a rabbet on the back side of the tails. This aids in laying out the pins, and it also relates to the cutting length of the router bit. More on this later. I stay away from the baseline with the router, then clean up the baseline with a chisel.

MARK THE PINS

Begin laying out the pins by smoothing the end grain of the pin board. I use a sharp block plane set for a light cut to remove the saw marks, then block sand a little with 220-grit sandpaper so I can make clean, crisp pencil lines. Make sure you keep the end flat and square. Next, place the tail board onto the end of the pin board, butting the shoulder of the rabbet on the back side tight to the inside face of the pin board. Using a 0.5 mm mechanical pencil, place the lead tight to the edge of the pins and draw a single line about ½" long alongside each of the four edges of the tails. Make sure you've marked right up to the edge of the tails. Don't let the body of the pencil push the lead away from the edge.

How to Make Condor Tails | **POWER-CUT DOVETAILS**

Mark the ends. Use a marking gauge to establish the length of the tails on the pin board.

Make a platform. A support platform helps to keep the router from tipping.

Just like when I hand-cut a dovetail, I scribe the position of the tails onto the end grain of the pin board with a marking knife. But unlike when making a hand-cut joint, I don't have the ability here to slightly shift the position of my backsaw to compensate for the thickness of the scribe line. Just as when chiseling the shoulder of the tail board, I want to be able to drop my chisel right into a scribe line to crisply establish the location of the pins. If I scribe with the tail board in one fixed position, I'll actually be marking outside the boundaries of the tail's socket, thus the socket will end up too large. So I'll need to shift the tail board left and right in order to get my scribe line exactly where I want it. The 0.5 mm pencil lines provide a way to observe minute movements of the tail board, which greatly helps in positioning the tail board for scribing.

The measurement of 0.5 mm is about 20 thousandths of an inch. I find that my marking knife, when used with light pressure, leaves a line about 10-thousandths wide. So in order to get that scribe line on the inside of the tail's socket, I'll need to shift the tail board over by about half the pencil line. This is where the pencil line is quite precise. You can tap the tail board over in small increments and watch as the tails begin to cover the pencil lines. You can adjust the offset by just a few thousandths with each tap, and more important, observe how much you're moving. For hardwoods such as the ash I am using, I'll move the tail board over so it covers about half the pencil line. For softer woods, or for joining a hardwood tail board to a softwood pin board, you can move farther. The more pencil line you cover, the tighter the joint will be.

THE DOVETAIL BOOK | 145

POWER-CUT DOVETAILS | How to Make Condor Tails

To scribe the tails onto the pin board, shift the tails to the right and observe the movement by watching how much of the pencil line gets covered by the tails. Now scribe just the left sides of each tail.

Now shift the tails to the left, using the same offset as before, and scribe the right sides of each tail. You've now placed the scribe lines precisely in the same plane as the sides of the tails, plus or minus a few thousandths.

Now use a marking gauge to define the length of the tails on the pin board. Close up the fence slightly (a few thousandths) so you end up with a nice tight fit here.

ROUT THIS WAY

Now that the position of the pins is established, the waste can be routed out. I usually just rest my router on the end of the pin board, but it's a good idea to make a platform around the pin board to help support the router. I position the top of the platform to be just a hair under the pin board surface. The platform is just there for security; you want the router to register off the top surface of the pin board. Use a marking gauge

Rout the waste. A trim router with a ¼" upcut spiral bit gets rid of most the initial waste.

Ready to chop. Get close with the router and the chiseling will be easier.

Chisel the pattern. Paring down establishes the final shape of the tail sockets.

Pattern maker. Clean work yields precise results.

146 | THE DOVETAIL BOOK

How to Make Condor Tails | POWER-CUT DOVETAILS

to define the rest of the tail sockets on the pin board.

Use a router to remove about ¼" depth of material from the waste areas. Stay about ¹⁄₃₂" away from the scribe lines.

This is where this technique really shines. Unless your pin board is cut from dead-perfect straight-grained stock, you'd never be able to pare down into this joint without splintering. After cutting the side that's with the grain, you might consider opening your own woodworking school. With the other side you'd be chucking the pin board through a picture window! And if you cut across the grain from the inside of the pin board, it would be difficult to keep a flat surface without lots of guide blocks and fussy setting. By paring out the last sliver of waste, only ¼" deep, you basically eliminate any grain direction issues. You also establish a nice exact pattern for the rest of the socket.

To pare out the waste, drop your chisel right into the scribed line all around and tap down. If you routed close, you don't even need a mallet. Be diligent here. This is the make-or-break moment. If you chop outside your scribe line, it will be glaringly obvious. Also, work carefully around the pin board now—those sharp arrises are easily damaged.

Here's the neat part. The pattern you just established will mean easy cutting of the rest of the socket using a top-bearing pattern bit. This is why we kept the base of the tails wider than the width of the router bit. The bit I'm using has a ⅝" cutting diameter and a 1" cutting length. You can easily get pattern bits in various configurations that work with this technique.

Because the bit is cutting parallel to the long grain, the router works easily, making long, straw-like chips. You can take a full-depth pass, although you do have to be aware of your feed rate. Take it slow so the bit doesn't get out of control. You might want to practice on some scrap. I like

Socket set. Set the router to mill the socket to full depth.

First pass. Get rid of most the waste on your first pass.

Second pass. Let the bearing rub the pattern for the final cut.

Corner chisel. You don't need a corner chisel to chisel the corners.

THE DOVETAIL BOOK | 147

POWER-CUT DOVETAILS | How to Make Condor Tails

The finished sockets. Everything is crisp, flat, and where it's supposed to be.

Easy now. Ease the back corners of the tails so they don't bruise the pins.

Like it grew that way. You'll know immediately if you've nailed it.

Pride of the flock. The rest of the world's dovetails will be jealous when you plane the finished joint.

to waste the bulk of the material first, staying away from the "pattern" at the top of the socket.

Once the majority of the waste is removed, I let the guide bearing lightly follow the pattern, just letting it kiss the chiseled surface. Do not be tempted to press the bearing hard to the pattern—you could dent it and ruin the fit of the joint. The router is removing very little material, so it's easy work, and you can use a light touch.

Use a chisel to clean the waste from the rounded areas in the corners where the bit couldn't reach. These are easy areas to work because they are below the surface; you don't have to fuss with making them perfect. Be careful to not split out the pin board. The platform here helps to prevent this.

To help ease the joint together, cut a chamfer on the back corners of the tails. I once cut the chamfer on the fronts of the tails. Once.

If you've done everything carefully, the joint will slide together sweetly, requiring just a few taps with a hammer. Don't fully seat the joint. If it's going in well at the beginning, the rest should be fine.

The final satisfaction comes after the glue dries. Use a sharp handplane to reveal the crisp details. ■

MAKE A DRAWER WITH A HALF-BLIND JIG

A dovetail jig enables anyone to make a classy drawer.

BY TOM CASPAR

As a woodworker, what's the first thing you notice when you open a drawer? The way it's put together, of course.

A cabinet drawer made with half-blind dovetails really stands out.

While a drawer that runs on metal slides doesn't need the strength of dovetailed joints, dovetails clearly say, "This drawer was built by a craftsman."

THE DOVETAIL BOOK | 149

POWER-CUT DOVETAILS | Make a Drawer with a Half-Blind Jig

In this article, you'll learn how to use a typical half-blind dovetail jig to make a standard drawer box. (It's called a "box" because the front of the drawer—what you see on the outside of the cabinet—is applied later.) Let's start with the anatomy of a box, then move on to setting up the jig.

MILL THE WOOD

Some species are better than others for making drawers. You'll want a wood that's relatively stable; once it has been milled, it should stay flat. And you'll want a wood that doesn't chip out when you rout it.

Maple and red oak are two excellent choices. We used maple.

If you have a table saw, jointer, and planer, it's best to mill the wood yourself in order to ensure that it's flat. (If the drawer pieces are cupped, the dovetails won't go together properly; if they're twisted, the drawer box will be twisted, too.)

Drawer boxes are typically made from ½"-thick material. Start with 4/4 (1") rough stock and saw it into individual drawer pieces (front, back, and sides). Cut the pieces about ¼" extra wide and 1" extra long. Be sure to mill some extra wood for testing the dovetail jig setups.

To ensure that your pieces stay flat, mill them down to ⅝" thick and let them sit for a few days. Joint the

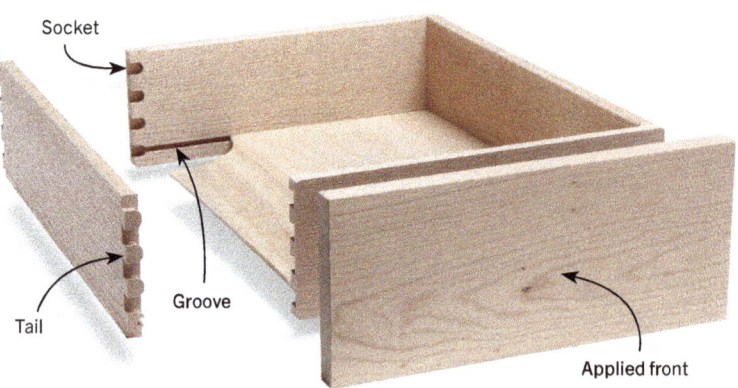

The box. Here's an exploded view of a typical drawer box. The front and back pieces have sockets cut into them; both are the same length. These pieces extend the full width of the drawer box.

The side pieces have tails formed on their ends. If all pieces are ½" thick—typical for a drawer box—the sides are cut ¼" shorter than the overall depth of the box.

A typical drawer box has an applied front and a ¼" plywood bottom trapped in a groove. The groove runs around all four sides of the box.

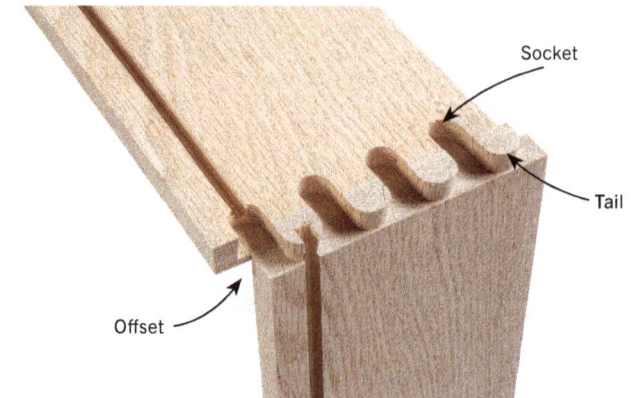

The joint. Here's what a typical half-blind dovetail joint looks like, unfolded. These boards are in the same orientation as they sit in the jig.

Tracing around the rounded fingers of the comb produces a series of sockets and tails. They're exactly the same width; when you assemble the two pieces, they'll automatically fit together.

The pieces will also align with each other, top and bottom, because they're offset by the right amount when you clamp them in the jig. The jig's stops create this offset.

Make a Drawer with a Half-Blind Jig | POWER-CUT DOVETAILS

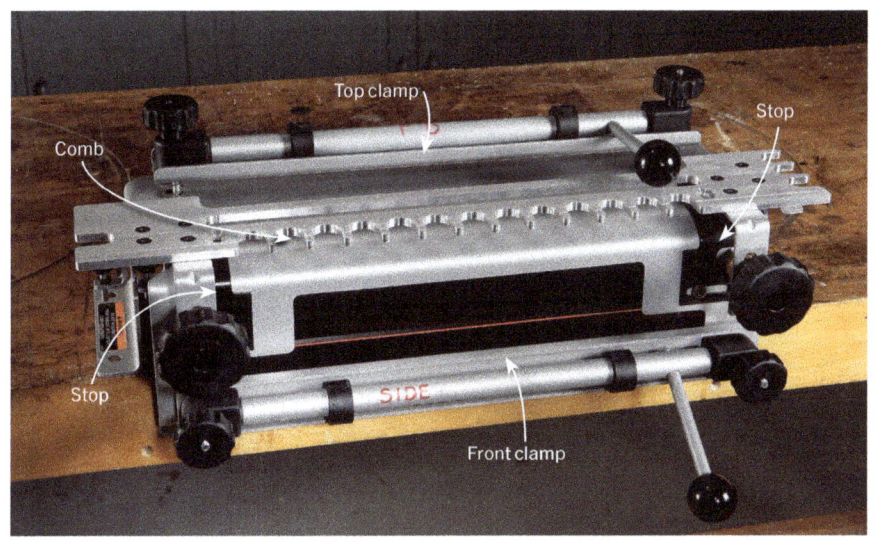

The jig. There are many jig manufacturers, such as CMT Orange Tools, Keller & Co., Leigh, and PORTER-CABLE. Most half-blind jigs have essentially the same parts. The most important one is the comb. Using a router equipped with a dovetail bit and a template guide, you trace around the comb to make the joint.

You always rout two boards at the same time. One is held horizontally by the top clamp; the other is held vertically by the front clamp. The clamps on this jig operate on a cam. Rotating each handle pushes a long clamping bar against the workpiece.

Most jigs have a pair of stops that position both workpieces left or right—relative to the comb.

YOUR JIG'S SPACING

On some jigs, the distance between the dovetails is 7/8"; on other jigs, it's 1". This is an important number when you're figuring out how wide the sides of your drawers should be. Why? Dovetail joints look best when they have half-pins at top and bottom. This rule limits the widths you can choose from. They will be increments of 7/8" or 1", depending on your jig, plus the spaces alloted for the half-pins.

Fortunately, when you're building drawer boxes that ride on slides and have applied fronts, the width of the drawer sides can be up to 1" less than the opening in the cabinet. For this model, we used a 1" jig and made the sides 4 1/4" wide. If you have a 7/8" jig, make the sides 4 1/2" high. In addition, make the bottom half-pin extra-wide, so the drawer-bottom groove runs through the center of the first socket.

7/8" spacing

1" spacing

THE DOVETAIL BOOK | 151

POWER-CUT DOVETAILS | Make a Drawer with a Half-Blind Jig

ADJUSTING YOUR ROUTER AND JIG

Every time you rout a new set of drawers, you'll make three adjustments to your router and jig. To test these adjustments, you'll need some scrap wood that is exactly the same thickness and width as the wood you'll be using for your drawer boxes.

1. FIT

First, install the correct template guide in your router's base and adjust the bit's depth of cut as specified in your jig's manual. (Typically, it's 5/8".) The depth of cut determines the joint's fit.

Install the jig's comb as specified in the manual and rout a pair of test pieces. Try assembling them.

If the fit is too loose, you'll see gaps between the two boards. Adjust the bit to cut about 1/64" deeper and try again.

If the fit is too tight, the joint won't go home. Adjust the bit to cut about 1/64" less deep and try again.

FIT

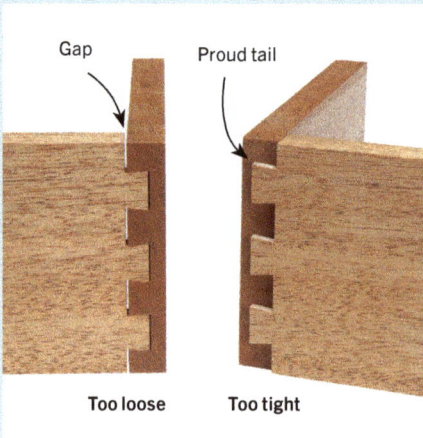

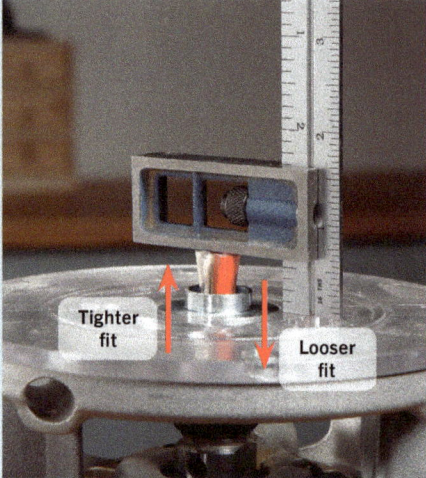

2. DEPTH

Next, adjust the position of the comb in or out. This affects how deep the sockets will be. Ideally, the sockets should be just deep enough so that the pins are flush when the joint is assembled.

In practice, it's best to position the comb so the sockets are about 1/64" too deep. This will compensate for any small error in your setup or minor variation in the thickness of the drawer's sides.

Adjusting nuts behind the comb's brackets determine the position of the comb. Both nuts must be the same distance from the front of the jig, so the comb remains parallel to the jig.

Examine the depth of the sockets on the test pieces you've made so far. If they're not deep enough (and this

DEPTH

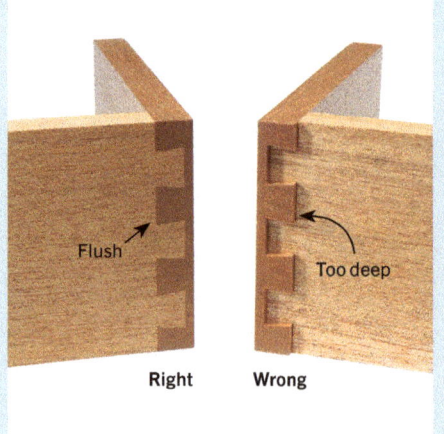

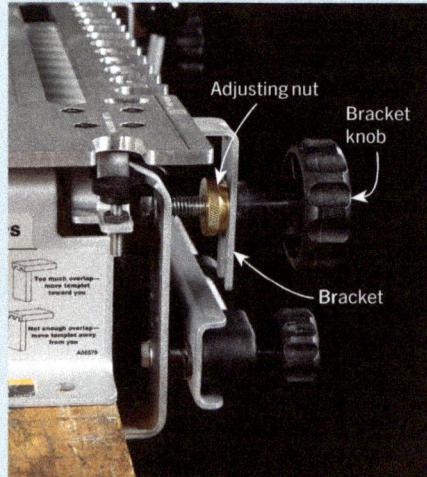

Make a Drawer with a Half-Blind Jig | **POWER-CUT DOVETAILS**

SPACING

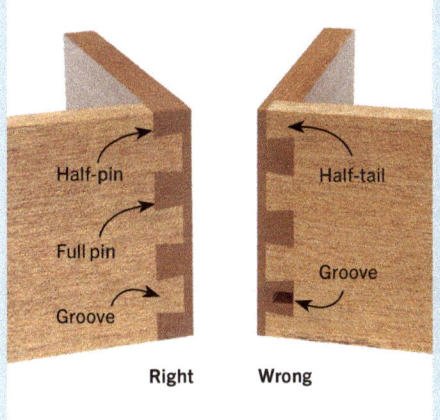

is easy to confuse with a joint that's too tight), turn each adjusting nut clockwise, closer to the jig, then tighten the bracket knobs. If the sockets are too deep—and this is the best place to start—loosen the knobs first, then turn the adjusting nuts counterclockwise, away from the jig. Re-tighten the knobs.

3. SPACING
Last, adjust the position of the stops on the jig. Moving them left or right affects how the joint looks. Your goal is to make a joint that begins and ends with half-pins rather than half-tails.

A half-pin doesn't have to be precisely half the width of a full pin, though. Close is good enough.

To set the stops, start with a new pair of test pieces. Determine where you want the groove for the drawer bottom to go, then draw "grooves" on the pieces. (Typically, grooves are located 3/8" to 1/2" above the bottom edge of a drawer.)

The groove should fall approximately in the middle of a socket, so a tail will cover it when the drawer is assembled.

To position each stop, loosen its adjusting screw and slide the stop all the way toward the end of the jig. Place one of the test pieces underneath the jig's comb and center the groove in the comb's first "U." Butt the stop up to the test piece and tighten it in place.

THE DOVETAIL BOOK | 153

POWER-CUT DOVETAILS | Make a Drawer with a Half-Blind Jig

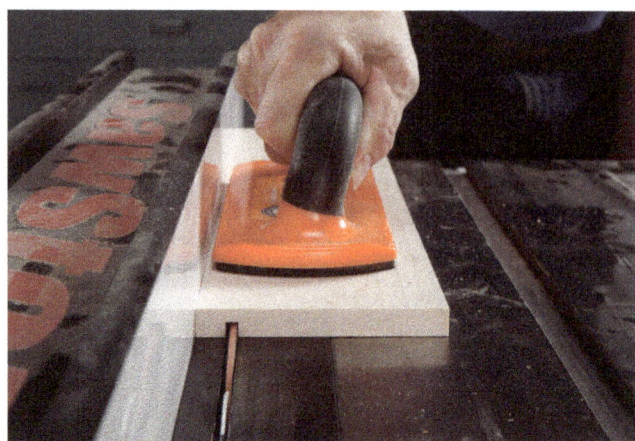

Cut all four drawer pieces to final size. Saw the drawer-bottom grooves. These grooves will help you position each piece in the dovetail jig.

Arrange the four parts of the drawer box. Mark them. ("FB" indicates a front/back board.) The left corners are routed on the jig's left side; the right corners are routed on the jig's right side.

Place two pieces in the left side of the jig. The drawer grooves face the jig's outer edge. Front/back boards go on top; side boards go in front. To avoid error, draw "FB" and "Side" on the jig, too.

Rout the dovetails. Remove these boards from the jig. Place the other two boards in the jig and rout them. After routing both left corners, move on to the right corners.

faces of the pieces and finish planing them down to ½". Joint the edges, rip the pieces to width, and crosscut them to final length.

SAW GROOVES FIRST

After all this work, the last thing you want to do is get confused about which board goes where on the jig. Here's an unorthodox marking system that's almost foolproof: Make the grooves for the drawer bottom before routing the dovetails. Once the grooves are cut, there's no mistaking which side of the board faces in and which side faces out on the jig—it's obvious.

Cut the grooves on the table saw using a standard blade (top left). Raise the blade ¼" high. You'll saw two overlapping cuts—start with the cut that's nearest the bottom edge.

Make a Drawer with a Half-Blind Jig | POWER-CUT DOVETAILS

Be sure to accommodate any under-mount hardware. Make this cut on all of your pieces. Adjust the saw's fence to make the second cut. Use a long piece of drawer-bottom plywood to test the groove's fit—you should be able to push the plywood along the groove with little effort.

Finally, mark each part of the drawer as a front, back, or side (p. 154, top right). You're ready to rout.

ROUT THE DOVETAILS

Use the test pieces to set up the jig (see pp. 152–153). It takes about a half-dozen tries to get everything right. Once you're set, you can rout many drawers without making any further adjustments as long as all the wood is the exact-same thickness.

Let's walk through how you'd make just one drawer. Place one of the drawer's front/back pieces and one of the side pieces in the left side of the jig (p. 154, bottom left). (The pieces are interchangeable, so it doesn't matter which two you use.) Make sure the boards are butted up to the stops and are flush with each other. It's a good idea to place a backer board in the jig, too. (For more tips, see pp. 157–160.)

Rout the dovetails (p. 154, bottom right). Place the router on the jig before you start it and turn it off before you remove it. Be careful not to tip the router. When you're done, inspect the joint before you remove the boards from the jig. If some portion is uncut, rout it again. Once the joint looks OK, remove the boards and place the other two front/back and side boards in the jig. Rout them as well. Repeat this procedure on the right side of the jig (top).

Place two boards in the right side of the jig. Once again, the grooves face the jig's outer edge. Rout these boards, then place the remaining two boards in the jig and rout them.

Assemble the drawer. Do not use glue. Measure the distance between the bottoms of the grooves using pinch sticks. Cut the drawer bottom $1/16$" smaller than this measurement.

Glue the drawer together. Assemble the front and sides first, then slide in the bottom. Add the back last. If the joints are nice and tight, you won't have to use clamps.

■ **POWER-CUT DOVETAILS** | Make a Drawer with a Half-Blind Jig

Square the drawer box. Clamp large L-shaped blocks to opposite corners. These blocks are made from two layers of MDF.

Round over. Complete all the edges of the drawer box—except the front. Support the router with a thick board that's the same width as the box.

GLUE THE DRAWER

Assemble the drawer—without glue—so you can figure out the exact size of the drawer-bottom (p. 155, center). While you could use a tape measure, there's no chance of making a math error if you use pinch sticks. (They're 1/8" thick and 3/4" wide.)

Here's how pinch sticks work: Insert one end of a stick in a groove, then extend the other stick until it bottoms out in the opposite groove. Clamp the sticks together, then rotate the assembly out of the grooves. Measure the length of the two sticks and subtract 1/16". Measure the drawer in the opposite direction and cut the drawer bottom to size. Sand all of the inside surfaces of the drawer and the top side of the drawer bottom.

Glue the drawer together (p. 155, bottom). You won't need a lot of glue—just apply it to the sockets with a small brush. Keep a damp rag handy for cleaning up the squeeze-out on the outside of the drawer. Make the joints as flush as you can.

After you add the last piece, make sure the drawer is square (top). Measure from corner to corner to see how close you are. The trick is bringing it in to square—L-shaped squaring blocks work well. After the glue dries, sand the joints flush.

Last, soften the edges of the drawer box with a 1/8" roundover bit (bottom). Before you start, put a big "X" on the front board—you don't want to round the front edges of this piece, where the applied front goes. Balance the router by using a board that's the same width as the drawer box. Use this piece to help rout both the inside and outside edges of each piece. ■

10 TIPS FOR USING A DOVETAIL JIG

A bunch of little tricks make the job go much easier!

BY TOM CASPAR

These tips will make using a dovetail router jig a snap.

1. PARAFFIN LUBE

When you rout a dovetail joint, your router should slide easily around the jig's fingers so you can feel when it's time to turn a corner. To eliminate drag, rub a piece of canning wax (paraffin) on top of the comb. You won't need much, but it sure helps!

2. CLIMB-CUT FIRST

To eliminate tear out inside a dovetail joint, make this your first step: Rout a shallow pass from right to left, all the way across the front board.

This scoring pass is a climb cut (routing in the direction of the bit's rotation). It goes in the opposite direction that you would normally move a router. Any climb cut presents a potential hazard, however: The bit can grab, suddenly pulling the router ahead. When you take a very shallow cut, though, that's usually not a problem.

When you've completed the scoring pass, rout the rest of the dovetail from left to right—in the opposite, and normal, direction.

Paraffin. A little wax keeps everything running smoothly.

Climb-cut first. A shallow climb cut helps to eliminate tear out.

POWER-CUT DOVETAILS | 10 Tips for Using a Dovetail Jig

Router nest. Make a nest to protect your router bit.

Bumper. A bumper protects your jig from accidental contact with the bit.

3. MAKE A ROUTER NEST

When you're constantly picking up and putting down a router with an exposed dovetail bit, try parking it on a platform. This way, the bit is protected and won't catch anything. Of course, you should still turn off the router first!

This platform is just a 1½" thick block the same size as the router's base, with ¼" x 1¾" sides nailed on all around. The hole is 1⅜" in diameter. Once the router is perched, the block's lip makes sure it stays put (see above).

Height block. Make measuring easy with a height block.

4. ADD BUMPERS

Accidentally routing into your dovetail jig will ruin your bit. On many jigs, the brackets that support the comb are directly in harm's way. If you're not sure of the router's position, you can chew right into them.

The solution is to add bumper blocks that prevent the router from getting too close to the brackets. On this jig, you can clamp the blocks directly to the comb; for models without extra-long combs, clamp tall blocks to the workbench.

5. MAKE A HEIGHT BLOCK

One of the most frustrating parts of setting up a dovetail jig is adjusting the bit to the correct depth of cut. Using a ruler can be very awkward. It's much easier to use a height block.

To make the block, drill two 5/32" holes all the way through a piece of plywood using a drill press (the holes must be precisely vertical). Turn a machine screw into each hole, then carefully adjust the height of each screw to match the correct depth of cut (usually ⅝"). Check both screws with a combination square. To ensure that the block sits flat, sand the heads of the screws while butting the block up against a larger chunk of wood. This keeps the block square.

Once you've made the block, adjust its screws until you get a tight-fitting joint. Store it with your jig. Next time around, adjusting the bit will be a cinch.

158 | THE DOVETAIL BOOK

10 Tips for Using a Dovetail Jig | **POWER-CUT DOVETAILS**

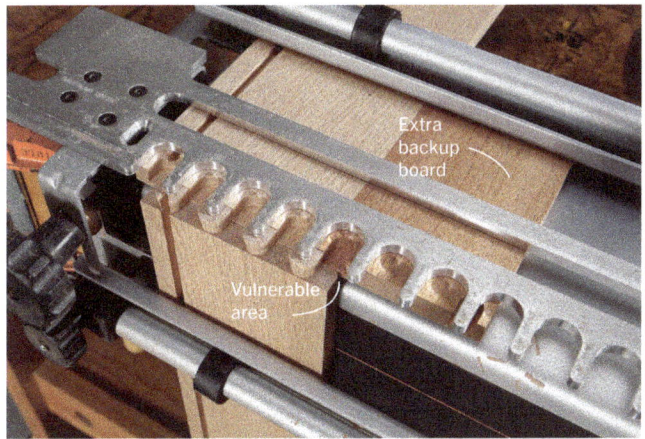

Disassembly. A soft backing surface will allow an unglued joint to be knocked apart without trying to hold the pieces in the air.

Backup board. Use a backup board to prevent errant routing of vulnerable areas.

6. ADD AN EXTRA BOARD

To prevent tear out, it's good practice to place an extra board on top of the jig next to the two boards you're routing.

Each time you rout a joint, rotate the backup board to a new corner. When you've used all four corners, cut off the ends of the board and start over.

Make a few backup boards when you mill the rest of the parts—they must be exactly the same thickness as your workpieces.

7. CUSHION THE BLOW

Here's a tip from the furniture-repair world. When you're knocking apart an unglued dovetail joint, place the pieces on material with some give, such as a towel, blanket, or router mat.

You don't have to hold the parts up in the air—the material will compress when you strike the wood, allowing the joint to slowly come apart.

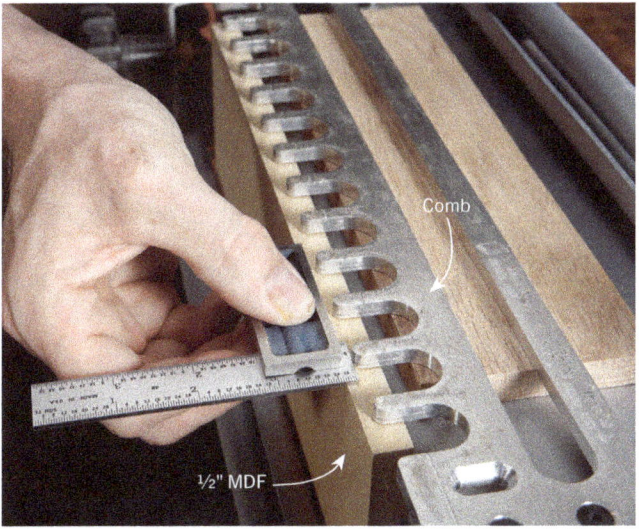

Register with a wide board. A wide board makes it easy to adjust the combs.

8. REGISTER WITH A WIDE BOARD

To control the depth of a dovetail joint, you move the comb in or out by adjusting the position of its brackets. When you're done adjusting each bracket, the comb must be perfectly parallel to the front of the jig. Here's an easy way to do that.

Clamp a 12"-wide piece of ½" MDF in the jig, as shown above. Position the comb so it's approximately in the correct position according to the jig's manual. Use a combination square to check both sides. Remove the MDF and clamp two boards in the left side of the jig. (They must be the same thickness as your final workpieces.) Make trial cuts and fine-tune the comb's position on the left side until the dovetails are exactly the correct depth. Put the MDF piece back in the jig, adjust the square to the comb's new setback on the left side, then slide the square to the right side of the comb. Adjust this side to match.

THE DOVETAIL BOOK | 159

POWER-CUT DOVETAILS | 10 Tips for Using a Dovetail Jig

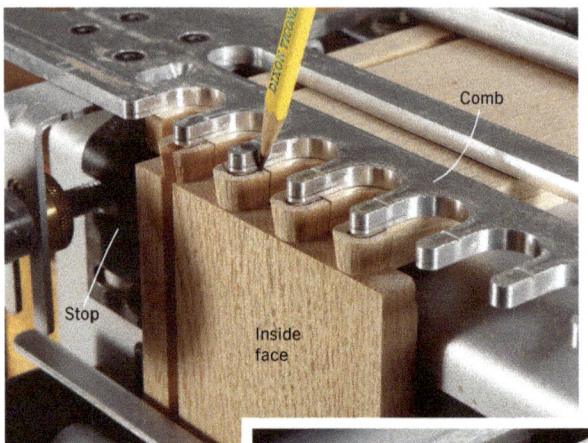

Set the stops. Trace the left side of the jig (left). Flip the board around and line it up on the right side of the jig (below).

9. SAME SPACING, LEFT AND RIGHT

Every dovetail jig has adjustable stops that set the distance between the first tail and the bottom edge of a workpiece. One stop is on the left side of the jig; the other stop is on the right side. Both stops have to be in the same position, relative to the comb's fingers, for the first-tail spacing to match.

Here's how to set them. First, make a test joint on the left side of the jig and adjust the left stop where you want it. Trace around the fingers of the comb (top left). Remove the board from the jig and flip it around so the opposite side faces out. Clamp the board in the right side of the jig so the tracings line up with the comb's fingers (center top). Butt the right stop up to the board. The spacing is now exactly the same on both sides of the jig.

10. SCORE VULNERABLE CORNERS

Tear out looks nasty on a well-made dovetail joint. Using a backer board can help reduce tear out, but it's not enough. An outside corner of the vertical piece may still chip out (bottom).

The best solution is to score this vulnerable area first, before routing. You could use a marking gauge, but if you don't have one, this little jig will do the trick. It works for boards mounted on either side of the jig—you just have to keep straight which corner to nick. ■

Prevent tear out. Tear out can occur in vulnerable corners (left). To prevent this, build a jig that shows you where to score the board (below).

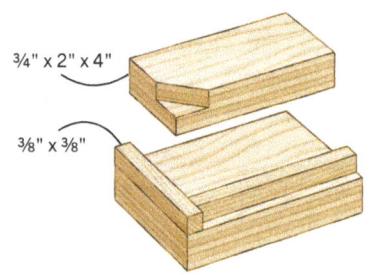

HALF-BLIND DOVETAILS BY JIG

Not everyone's ready to tackle hand-cut dovetails; here's how to get the most from your router and jig.

BY BILL HYLTON

Dovetails are prime joints. Long history, great appearance, and cachet; used in boxes, drawers, and carcases. But for many woodworkers, cutting dovetails the traditional way—with saw and chisels—is an insurmountable challenge.

If you aren't ready to tackle hand-cut dovetails, there are plenty of router accessories on the market to help. There are so many in fact, and they have so many variations in setup and operation, that I'm going to narrow my focus to the most common: the half-blind dovetail jig.

The typical half-blind dovetail jig consists of a metal base with two clamping bars to hold the workpieces. A comb-like template rests on the top to guide the router in cutting both pieces at once. The appropriate bit and bushing are packaged with the jig. Usually you use a ½" 14° dovetail bit and a ⁷⁄₁₆" guide bushing to make the cuts.

Use any router, which is to say, the one you have. I typically use a 2-horsepower fixed-base model. The ability to plunge is irrelevant, and plunge routers generally are awkward for work on the edge owing to their high centers of gravity. Brute

POWER-CUT DOVETAILS | Half-Blind Dovetails by Jig

power doesn't contribute anything. When the urge to rout half-blind dovetails seizes you, get out your jig and clamp it at the edge of your workbench. Presumably, you'll have stored the instructions and the right bit and guide bushing with the jig.

Select your materials and make sure all like parts are jointed and planed uniformly. Not all the parts must be the same thickness. The fronts can be ¾" thick, and the sides and backs ½" thick, for example. Or ¾" and ⅝". Everything can be 11/16". Just be certain the fronts are consistently sized, the sides are consistently sized, and also the backs.

Setting the depth of cut. You must account for the template thickness as well as the cut itself. Use a small machinist's square to set the bit extension from the baseplate.

Line up the workpieces carefully. As you clamp the pieces in the jig, snug the end of the socket piece (it's on top) against the inner face of the tail piece (it's on the front). Make sure the pieces are flush against the guide pins or stops. (The pin on the jig is hidden by the socket board and clamping bar.)

SET UP THE ROUTER

Install the guide bushing. (If you have a centering mandrel, use it to center the bushing to the bit's axis.) Adjust the router so the collet is relatively close to the bushing. Carefully insert the dovetail bit through the bushing and into the collet. Tighten the collet nut.

Adjust the depth of cut next, as shown on p. 163. When you do this, turn the bit slowly by hand to absolutely ensure that the bit doesn't contact the bushing. The cutting end of the bit is too large to pass through the bushing. If you use a steel bushing, it will damage the bit's carbide, so you want to avoid accidental contact.

Check your jig's instructions for the recommended depth-of-cut setting. It's often in the 21/32" to 23/32" range, depending upon the thickness of the template. A good generic starting point is ⅜" plus the template thickness (to get an accurate measurement of the template, use dial calipers).

Prevent chipping. Chipping along the shoulder of the tail piece is a problem. To eliminate this, make a shallow scoring cut across the tail piece first. A climb cut—where you feed the router from right to left—is most effective here. Just be sure the router doesn't get away from you.

Cut. Move the router along the template, feeding the router into each slot and keeping the guide tight against the template as you come out of one slot and round the finger into the next slot. Any little bump on either tail or socket will prevent assembly of the joint.

CLAMP THE WORK IN THE JIG

The work has to be clamped in the jig in a particular way. When you cut following the template, tails are formed on the front board in the jig, and sockets into which the tails nest are cut simultaneously into the top board.

So the socket piece—and that's always the drawer front or back—is on top. The tail piece—the drawer side—is at the front. Alignment is critical: The tail board overlaps the end of the socket board, and its end must be flush with the upper face of the socket board. The boards must be perpendicular to each other. In addition, the tail board is offset. Both boards are clamped in the jig with their "inside" faces out.

Here's an easy way to do it. Roughly position the tail piece in the jig, with its top end well above the jig. Slip the socket piece under the top clamping bar, and butt it tightly against the tail piece. Clamp it firmly. Now loosen the clamp holding the tail piece and lower it until its end is flush with the other workpiece. Clamp it firmly.

Both pieces need to be against the alignment pins or stops. These pins align the parts so they are offset exactly 7/16", which is half the center-to-center spacing of standard router-cut half-blind dovetails (7/8"). Every jig has these pins on the right and on the left. Use those on the left for now.

The template must rest flat on the work. Its fore-and-aft alignment is critical to the fit of the joint, but don't worry about it for now. Use the out-of-the-box setting for your initial test cuts, and adjust as necessary.

CUT A TEST JOINT

Rest the router on the template with its bit clear of the work. Switch on the router and make a quick, shallow scoring cut across the tail piece, feeding from right to left (yes, this is a climb cut).

The purpose of this cut is to prevent tear out along what will be the inside shoulder. What often happens is that the bit blows out splinters as it emerges from each slot of the jig's template. If there's no shoulder established first, these splinters can run down the face of the drawer side, defacing it.

Now rout the dovetails, slot by slot, beginning on the left and working to the right. Feed the router into each slot of the template, then back it out. Keep the router firmly against the template as you round the tip of each template finger; you want to completely form each tail—no little lumps.

I usually zip back through the slots after the first pass, just to be sure I didn't pull out of a slot too soon, leaving that socket only partially cut. Don't just lift the router from the template. The bit will ruin both the cut and the template. Instead, turn off the power and pull the router toward you, getting it well clear of the jig before lifting it.

Take a good look at the work and be sure you haven't missed a spot. If you have, re-rout it before moving anything clamped in the jig. Remove the template, unclamp the work, and test assemble the joint.

FINE-TUNE THE SETUP

Slip the test pieces together. Maybe something's not quite right. Perhaps the fit is too loose. Or too tight. Or the sockets aren't deep enough. Or the parts are a little offset. All of these ills are cured with some fine-tuning.

DETAIL: SETTING DEPTH OF CUT

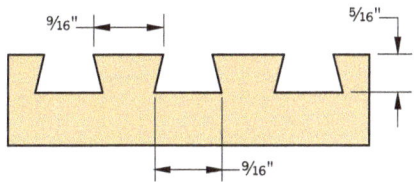

Optimum. The pin formed matches the slot cut by the dovetail bit.

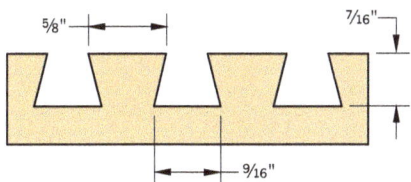

Too deep. The pin formed is wider than the slot.

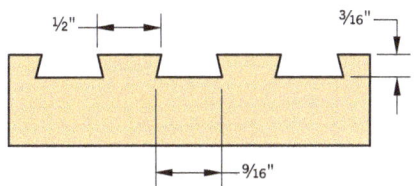

Too shallow. The pin formed is narrower than the slot.

POWER-CUT DOVETAILS | Half-Blind Dovetails by Jig

Misfits. If your test joint doesn't fit exactly, the nature of the misfit shows you how to correct it on your next cut. If the tails are tight or loose in the sockets (top), adjust the cutting depth. If the tails fit, but aren't flush (bottom), adjust the position of the template.

The bit's cut depth is the primary control of fit. The way it works is shown in the drawing, "Setting Depth of Cut," p. 163. The cut—the socket—is always the size of the cutter. But when you alter the depth of the cut, the width of the material left between sockets changes. Because you are cutting both tails and sockets at the same time, the material between the sockets is in fact the tail.

In practice, this aspect of the setup is at the same time deceptive and frustrating. The transition from "no fit" to "perfect fit" is abrupt—just a 1/32" change can make all the difference. What often happens is that you lose confidence in the adjustment regimen after one or two incremental changes with no apparent effect. "Well, this isn't getting me anywhere!" you think, and start adjusting in the other direction. And you seesaw between increasing and decreasing the cut depth, never hit the right setting, get totally frustrated, and shelve the jig, never to use it again.

Take heart. Remember that woodworkers have been using these jigs for decades, and that routers have been pretty primitive tools for most of that time. You can do it. Be patient, methodical, and persistent. Here's what you do:

- Reduce the cut depth to loosen the fit.
- Increase cut depth to tighten the fit.

Once the depth of cut is dead on, analyze a new test cut and determine if other adjustments are needed.

The relationship of the joint surfaces is controlled by the template's fore-and-aft position. Ideally, the surfaces are flush when the joint is seated tightly.

- If the side is recessed, the pin is short and the socket is long. Shift the template back.
- If the side is proud of the front's end, the pin is long and the socket is short. Shift the template forward.

Your jig's instruction sheet should explain exactly how to accomplish this. Generally, the template bracket sits against a nut on the mounting stud. Turn the nut and the template moves. These studs usually are 1/4"-20 bolts, so a full turn of the nut will move the template in or out 50 thousandths of an inch.

Look at the edges next. When the joint is assembled, the adjoining edges should be flush. If they aren't, you may not have had the workpieces snug against the alignment pins. Or the pins may be slightly misadjusted.

Any other problems you have will have stemmed from misalignment of the workpieces in the jig. Make sure the top surface of the socket piece is flush with the top end of the tail piece, that they are at right angles to each other, that the template is square to the workpieces, and so forth.

When you've successfully fine-tuned the setup using the alignment pins on the left, cut a test joint at the other end of the jig. Do any additional tuning needed there.

DOVETAILING THE GOOD WOOD

Before starting on the actual project parts, make sure you're organized. The parts are worked "inside out." If you are doing drawers, the sides always go on the front of the jig, and the fronts and backs always go on the

Half-Blind Dovetails by Jig | POWER-CUT DOVETAILS

top. Some joints are cut on the right side of the jig, others on the left. It's easy to get mixed up, whether you're dovetailing one drawer or 50.

A good way to avoid confusion is to label the parts on what will be their inside faces, as shown in below. Where you put the labels is as important as what they are. The letters are always associated with a particular part. Put the letter at the bottom so you know which edge goes against the alignment pins. On the jig itself, mark the two-letter combinations beside each pair of alignment pins, as indicated in the drawing. As you clamp the parts into the jig, orient the letters toward the pins, and check the combination. If it isn't on your list of two, you are at the wrong end of the jig. ∎

DETAIL: SAMPLE SETUP

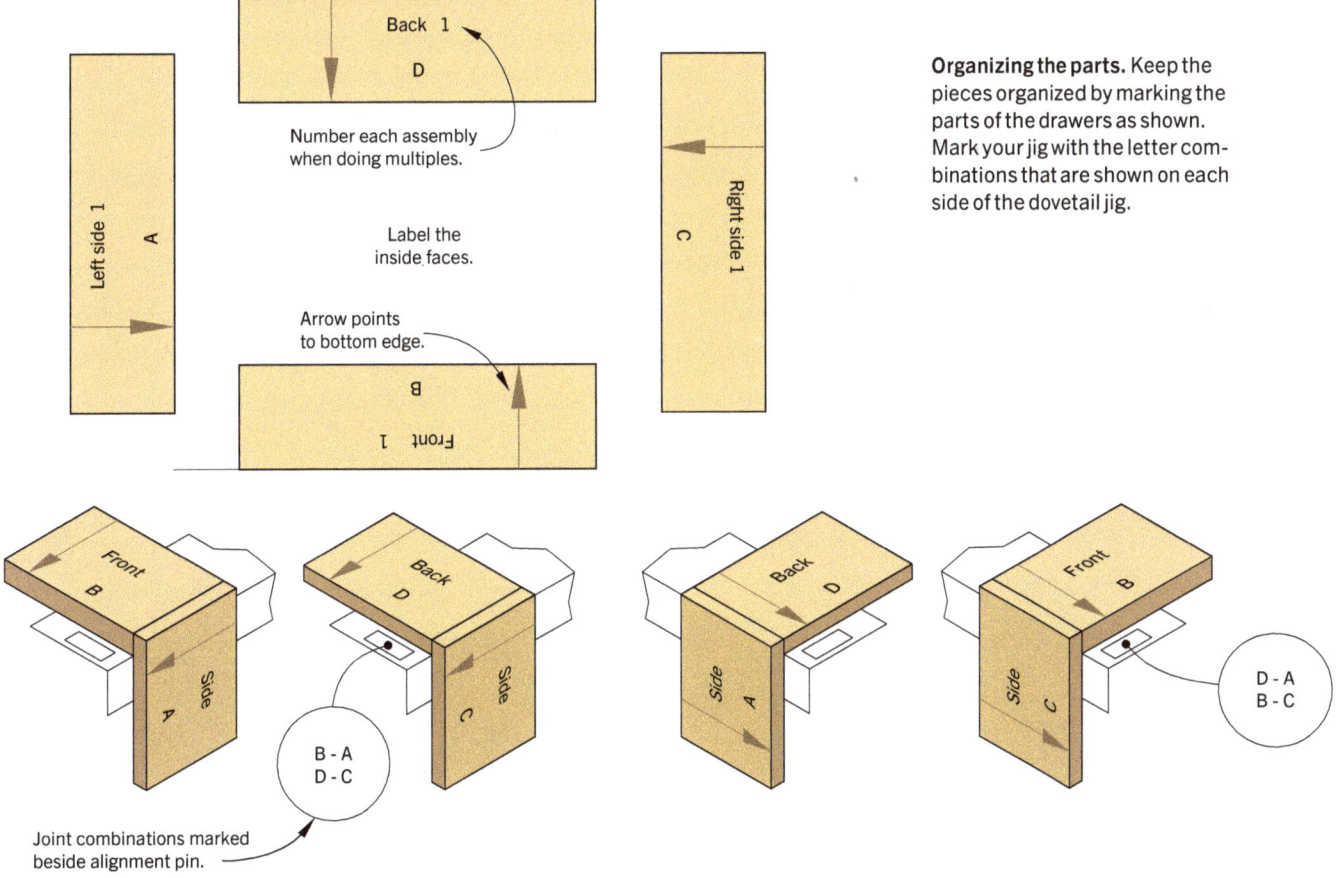

Organizing the parts. Keep the pieces organized by marking the parts of the drawers as shown. Mark your jig with the letter combinations that are shown on each side of the dovetail jig.

CONTRIBUTORS

Tools for Dovetailing (p. 6): Jason Zentner, Photog.

Your First Hand-Cut Dovetails (p. 10): Al Parrish, Photog.

4 Ways to Make Dovetailing Easier (p. 19): Hayes Shanesy, Illus.

Frank Klausz on Learning to Cut Dovetails (p. 20): Al Parrish, Photog.; Matt Bantly, Illus.

Dovetail Ruler Trick (p. 28): Christopher Schwarz, Photog.

Dovetail Dilemma (p. 31): Father John Abraham & Narayan Nayar, Photog.

Precise Hand-Cut Dovetails (p. 36): Joe Gohman, Illus.; Ramon Moreno, Photog.; Vern Johnson, Art Director

Tips to Avoid Dovetail Gaps (p. 48): Al Parrish, Photog.

Houndstooth Dovetails (p. 50): Al Parrish & Wolfgang Busse, Photog.

Tapered Sliding Dovetails (p. 67): Ben Owen, Photog.

Compound-Angle Dovetails (p. 71): Chad Stanton & Jason Zentner, Photog.; Frank Rohrbach, Illus.

Impossible Dovetails (p. 75): Frank Rohrbach, Illus.; Tim Johnson, Ed.

Mitered Dovetail Box (p. 82): Christopher Walker, Photog.

The Telegraphing Effect (p. 89): Martin Greshoff, Photog.

Dovetail Station (p. 90): Frank Rohrbach, Illus.; Jason Zentner, Photog.

Shaker Tray with a Little Embellishment (p. 98): Logan Wittmer, Photog.

Bandsawn Dovetails (p. 108): Jason Zentner, Photog.

Power-Assisted Half-Blind Dovetails (p. 113): Scott Gibson, Photog.

Router Table Dovetails (p. 120): Frank Rohrbach, Illus.; Jason Zentner, Photog.; Tom Caspar, Ed.

Sliding Dovetails with a Router (p. 127): Al Parrish, Photog.

Make a Sliding Dovetail at the Table (p. 131): Chad Stanton & Jason Zentner, Photog.; Frank Rohrbach, Illus.; Tom Caspar, Ed.

Table Saw Dovetails (p. 136): Frank Rohrbach, Illus.; Jim Stack, Photog.; Tim Johnson, Ed.

How to Make Condor Tails (p. 141): Father John Abraham, Photog.

Make a Drawer with a Half-Blind Jig (p. 149): Frank Rohrbach, Illus.; Jason Zentner, Photog.

10 Tips for Using a Dovetail Jig (p. 157): Frank Rohrbach, Illus.; Jason Zentner, Photog.

Half-Blind Dovetails by Jig (p. 161): Bill Hylton, Photog.

MANUFACTURERS

Visit your local woodworking store or look online for these brands mentioned in the book.

Amana Tool (router bits) amanatool.com

Calvo (bronze woodcarving mallet) davidcalvo.com

CMT Orange Tools (router jigs) cmtorangetools.com

DESTACO (horizontal toggle clamps) destaco.com

Freud (dovetail router bit) freudtools.com

Keller & Co. (router jigs) kellerdovetail.com

Leigh (router jigs) leightools.com

MLCS (router bits) mlcswoodworking.com

OLSEN (Cool Blocks for bandsaws) olsonsaw.net

Pfeil (chisels) pfeiltools.ch

PORTER-CABLE (router jigs) portercable.com

Robert Sorby (chisels) robert-sorby.co.uk

Stanley (chisels) stanleytools.com

METRIC CONVERSIONS

In this book, we've used inches, yards, ounces, and pounds, showing anything less than one as a fraction. If you want to convert those to metric measurements, please use the following formulas:

Fractions to Decimals

$1/32$ = .03125 $1/4$ = .25

$1/16$ = .0625 $1/2$ = .5

$1/8$ = .125

Imperial to Metric Conversion

Length

Multiply inches by 25.4 to get millimeters

Multiply inches by 2.54 to get centimeters

Multiply yards by .9144 to get meters

For example, if you wanted to convert $1 1/8$ inches to millimeters:

1.125 in. x 25.4 mm = 28.575 mm

And to convert $2 1/2$ yards to meters:

2.5 yd. x .9144 m = 2.286 m

Weight

Multiply ounces by 28.35 to get grams

Multiply pounds by .45 to get kilograms

For example, if you wanted to convert 5 ounces to grams:

5 oz. x 28.35 g = 141.75 g

And to convert 2 pounds to kilograms:

2 lb. x .45 kg = .9 kg

INDEX

adjusting your router and jig 152–153
a dovetail a day 18
Arts & Crafts 32
assemble 17, 77, 79, 81, 94, 103, 117, 119, 126, 155, 156

bandsawn dovetails 108–112
box project 82–88

chamfer 56, 95, 97, 104, 148
chisel 25, 26, 54, 94, 102, 144, 146
chisel modification 12
chop the shoulder 44
climb-cut first 157
compound-angle dovetails 71–74
compression 48, 49
condor tails 141–148
coping saw 8
cord wrap 104
corner clamps 19
cut by eye 34
cutting gauge 7

darken the lines 19
dovetail marker 12, 21, 53, 57, 68, 84
dovetail station 90–97
Dozuki saw 7, 77
drawers 3, 60, 61
drawer slips 64
dry fit, don't 58

Federal 32
five Ps of sawing 57

half-blind dovetails 113
half-blind dovetails by jig 161–165
half-blind, fake 49
half-blind jig 149–156
half pin 23, 24, 26, 34, 51, 54, 73, 111, 115, 116, 138, 139
houndstooth dovetail 50–58

impossible dovetails 75–81

jeweler's saw 63
jig 8, 11, 36, 67, 107, 109, 120, 127, 128, 129, 149, 160

kerfs 14
Klausz, Frank 3, 5, 21, 27, 31, 34, 141, 166

mallet 8, 11–17, 35, 37, 41, 111, 114, 116, 118, 119, 126, 147, 166
marking knife 19
mitered dovetail 82

paraffin lube 157
Pope's coffin 27
puzzles 75
rabbet 28, 29, 51, 55, 65, 100, 134, 139, 144
rabbet trick 28
remove waste 15, 140
rip teeth 19
router 19, 23, 48, 69, 70, 92, 102, 103, 107, 113, 142, 161, 166
router plane 69
router table 107, 120, 129–135
router table dovetails 120–126
rule of halves 34
ruler 28, 29, 30, 143, 158

saddle square 6
Shaker tray project 98–105
shims 8, 93, 135, 138
sizing pins and tails 31
sliding dovetail 131–135
sliding dovetails 127–130
sliding T-bevel 6
spiral up-cut bit 117
square 6, 27, 122, 124, 140, 156
strength 9
striking knife 7
strop 8
table saw 38, 47, 64, 78, 84, 93, 107, 109, 131, 137, 150, 154
table saw dovetails 136–140
tapered sliding dovetails 67–69
telegraphing effect 89
through-dovetail 11, 50, 51, 83, 130, 137
tips for using a jig 157–160
tools, cutting and chopping 7
tools, layout 6, 12

Underhill, Roy 31

wider board 22
wood selection 37, 61

Text and photographs © 2022

All rights reserved. Excepting patterns, no part of this book may be reproduced or transmitted in any form or by any means, electric or mechanical, including photocopying, recording, or by any information storage and retrieval system, without written permission from the Publisher.

Readers may make copies of patterns for personal use. The patterns themselves, however, are not to be duplicated for resale or distribution under any circumstances. Any such copying is a violation of copyright law. All text and photos previously published in *Popular Woodworking*.

Publisher: Paul McGahren
Editor: Kerri Grzybicki
Design & Layout: Clare Finney

Cedar Lane Press
PO Box 5424
Lancaster, PA 17606-5424

Paperback ISBN: 978-1-950934-94-2
ePub ISBN: 978-1-950934-95-9

Library of Congress Control Number: 2022945144

Printed in the United States of America
10 9 8 7 6 5 4 3 2 1

Note: The following list contains names used in *The Dovetail Book* that may be registered with the United States Copyright Office:

Amana Tool; *America;* CMT Orange Tools; DESTACO; DEWALT; Freud; Gorilla Glue (Gorilla Tape); *Heirs of the Fisherman: Behind the Scenes of Papal Death and Succession;* Honda (Civic); Keller & Co.; Leigh; Liquid Paper; MLCS; NPR; OLSEN (Cool Blocks); Pfeil; PORTER-CABLE; Robert Sorby; Stanley Tools; The Beatles; The Catholic News Service; *The Deaths of the Popes; The Encyclopedia of Furniture Making;* and *The Workbench Book.*

The information in this book is given in good faith; however, no warranty is given, nor are results guaranteed. Woodworking is inherently dangerous. Your safety is your responsibility. Neither Cedar Lane Press nor the author assume any responsibility for any injuries or accidents.

To learn more about Cedar Lane Press books, or to find a retailer near you, email info@cedarlanepress.com or visit us at www.cedarlanepress.com.

Popular Woodworking Shop

STORE • PROJECTS • TECHNIQUES • TOOLS • VIDEOS • BLOG • SUBSCRIBE • MY ACCOUNT

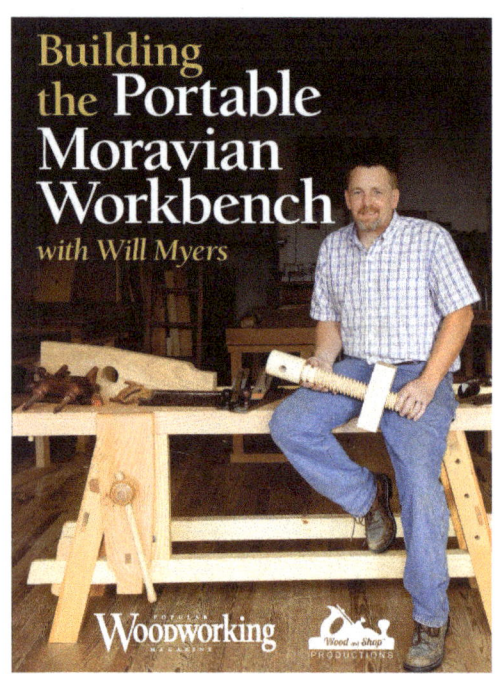

Building the Portable Moravian Workbench
Video Download featuring Will Myers

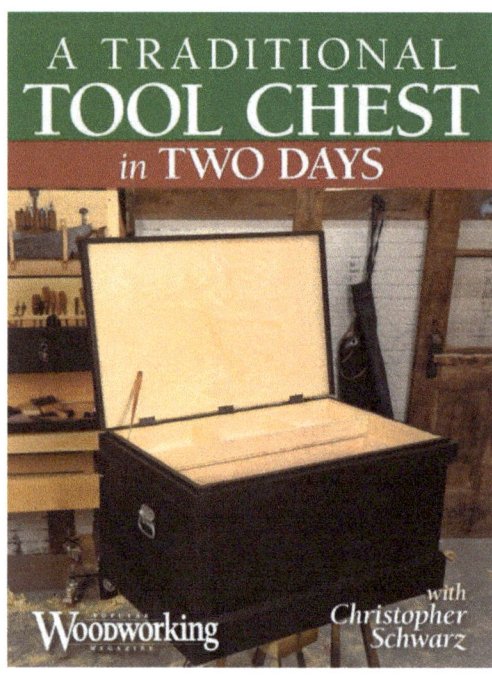

A Traditional Tool Chest in Two Days
Video Download featuring Christopher Schwarz

popularwoodworking.com/store

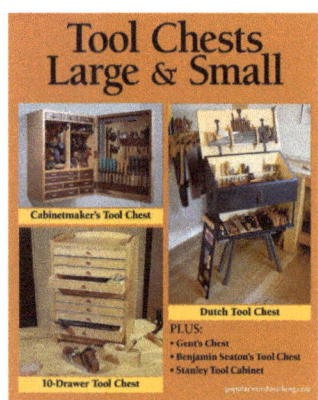

Six Tool Chests Large & Small
Project Download

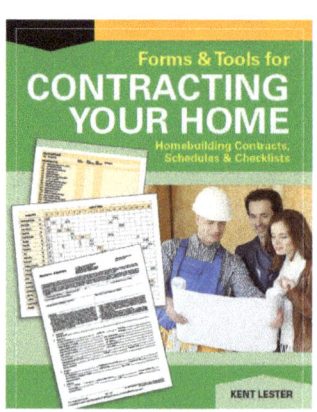

Forms & Tools for Contracting Your Home
eBook by Kent Lester

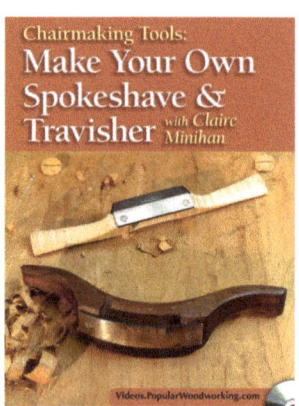

Make Your Own Spokeshave & Travisher
Video Download with Claire Minihan

www.ingramcontent.com/pod-product-compliance
Lightning Source LLC
Chambersburg PA
CBHW040932240426
43673CB00051B/1959